# Abel's Proof

An Essay on the
Sources and Meaning
of Mathematical
Unsolvability

Peter Pesic

The MIT Press
Cambridge, Massachusetts
London, England

This book was set in Palatino by Interactive Composition Corporation and was printed and bound in the United States of America.

Permission for the use of the figures has been kindly given by the following: Dover Publications (figures 2.1, 3.1); Conservation Departmentale de Musées de Vendée (figure 2.2); Department of Mathematics, University of Oslo, Norway (figure 6.1); Francisco González de Posada (figure 10.1, which appeared in *Investigación y Ciencia*, July 1990, page 82); Robert W. Gray (figure 8.3); Lucent Technologies, Inc./Bell Labs (figure 10.2); National Library of Norway, Oslo Division (figure 10.3).

I am grateful to Jean Buck (Wolfram Research), Peter M. Busichio and Edward J. Eckert (Bell Labs/Lucent Technology), Judy Feldmann, Chryseis Fox, John Grafton (Dover Publications), Thomas Hull, Nils Klitkou (National Library of Norway, Oslo Division), Purificación Mayoral (*Investigación y Ciencia*), George Nichols, Lisa Reeve, Yngvar Reichelt (Department of Mathematics, University of Oslo), Mary Reilly, and Ssu Weng for their help with the figures. Special thanks to Wan-go Weng for his calligraphy of the Chinese character *ssu* (meaning "thought") on the dedication page.

Library of Congress Cataloging-in-Publication Data

Pesic, Peter.
  Abel's proof: an essay on the sources and meaning of mathematical unsolvability / Peter Pesic.
     p.   cm.
  Includes bibliographical references and index.
  ISBN 0-262-16216-4 (hc : alk. paper), 0-262-66182-9 (pb)
  1. Equations, roots of. 2. Abel, Niels Henrik, 1802–1829. I. Title.

QA212 .P47   2003
512.9′4—dc21                                            2002031991

10  9  8  7  6  5

for Ssu,
thoughtful and beloved

# Contents

Introduction    1

1  The Scandal of the Irrational    5

2  Controversy and Coefficients    23

3  Impossibilities and Imaginaries    47

4  Spirals and Seashores    59

5  Premonitions and Permutations    73

6  Abel's Proof    85

7  Abel and Galois    95

8  Seeing Symmetries    111

9  The Order of Things    131

10  Solving the Unsolvable    145

Appendix A: Abel's 1824 Paper     155
Appendix B: Abel on the General Form of an Algebraic
    Solution     171
Appendix C: Cauchy's Theorem on Permutations     175
Notes     181
Acknowledgments     203
Index     205

# Introduction

In 1824, a young Norwegian named Niels Henrik Abel published a small pamphlet announcing a new mathematical proof for an old problem. Few read his pamphlet or paid much attention. Five years later, Abel died at age twenty-six, just before his work started to receive wide acclaim. Developed by others, his insights became a cornerstone of modern mathematics. Yet, apart from mathematicians, his ideas remain generally unknown.

This book tells the story of Abel's problem and his proof. The text contains a minimum of equations, so that the main lines of the argument should be accessible to people who are intrigued by ideas but feel uncomfortable with mathematical detail. Boxes develop arguments and give examples in greater depth, but these can be skipped without guilt. The appendixes go further still, including an annotated translation of Abel's pamphlet. The notes give references and suggestions for further reading.

Abel's proof concerns the solution of algebraic equations, that is, equations of the form $a_n x^n + a_{n-1} x^{n-1} + \cdots + a_0 = 0$, where $x$ is the "variable" whose "solutions" or "roots" we are seeking and $a_n, a_{n-1}, \ldots, a_0$ are constant "coefficients." We classify algebraic equations by $n$, the highest power. If $n = 1$, the equation is linear and $x$ has one root. If $n = 2$, the equation is quadratic, and we are taught in high school that the general solutions for any quadratic equation $a_2 x^2 + a_1 x + a_0 = 0$ are given by $x = \frac{-a_1 \pm \sqrt{a_1^2 - 4a_2 a_0}}{2a_2}$, no matter what integers or fractions $a_2, a_1, a_0$ might be. The quadratic equation is thus "solvable in radicals," as mathematicians say, meaning that its solutions can be expressed in terms of radicals (in this case, the square root, but in general meaning any root, such as a cube root, a fifth root, etc.), combined with the four ordinary algebraic operations (addition, subtraction, multiplication, division). This is as far as one goes in high school, except perhaps to learn that there are equations of still higher degree: if $n = 3$, we get equations like $x^3 + 2x^2 + x - 1 = 0$, called cubic or of the third degree. Likewise, there are fourth-degree equations (also called quartic) that involve $x^4$, fifth-degree equations (called quintic) that involve $x^5$, and so on, so that an $n$th-degree equation has powers of $x$ up to and including $x^n$.

For many people, the quadratic formula is a dim memory at best, though most of us recall that a quadratic equation has solutions that can be found from a certain formula. What about cubic equations or quartic equations? It turns out that they are also solvable in radicals. Their solutions are more complex than the quadratic formula, but they are available in books and used to be studied routinely in school texts of the nineteenth century. In the case of the cubic equation, the solutions involve cube roots of square roots. It all seems very routine, even tedious; we expect that all equations have solutions, no matter what their degree, but that the solutions get more and more complex the higher the degree. So far, this all appears very tame.

But there is a great surprise. In general, a quintic equation is *not* solvable in radicals. That is, although there are certain special quintic equations that have solutions we can express in terms of radicals, if we take the general equation of the fifth degree, $a_5x^5 + a_4x^4 + a_3x^3 + a_2x^2 + a_1x + a_0 = 0$, there are infinitely many values of the coefficients for which there is no way of expressing what values of $x$ would solve this equation in terms of some formula involving a finite number of square roots, cube roots, fifth roots, and other algebraic expressions. That is, there *are* values of $x$ that will satisfy the given quintic equation, but we cannot express them in any finite formula, however complex, involving just roots and powers, added, subtracted, multiplied, and divided in some way, as we could for all equations up to the quintic. What is worse, the same holds for equations of any degree higher than five: in general, equations of the sixth, seventh, eighth, . . . , $n$th degree are not solvable in radicals either.

Why? What happened to the pattern of algebraic equations each having solutions? What is it about the fifth degree that causes the problem? Why does it then go on to affect all higher degrees of equations? Most of all, what is the significance of this breakdown, if one can use such a word?

Such questions have moved me since childhood. Mathematical symbols may indicate hidden truths that have deep human significance, even as they transcend the human. Abel's proof contains a prime secret: how can a search for solutions yield the unsolvable? Perhaps if I tried hard enough, I could understand. I studied modern texts, but the key remained elusive. Absorbed in advanced studies, experts may cease to wonder about the elementary. They might not notice the kind of basic insight I was seeking. To find it, I needed to return to the sources, retracing the journey recounted in this book. The story begins in ancient Greece and has climactic scenes in Norway and France in the 1820s. What Abel found is indeed surprising and strangely beautiful.

# 1

# The Scandal of the Irrational

The story begins with a secret and a scandal. About 2,500 years ago, in Greece, a philosopher named Pythagoras and his followers adopted the motto "All is number." The Pythagorean brotherhood discovered many important mathematical truths and explored the ways they were manifest in the world. But they also wrapped themselves in mystery, considering themselves guardians of the secrets of mathematics from the profane world. Because of their secrecy, many details of their work are lost, and even the degree to which they were indebted to prior discoveries made in Mesopotamia and Egypt remains obscure.

Those who followed looked back on the Pythagoreans as the source of mathematics. Euclid's masterful compilation, *The Elements*, written several hundred years later, includes Pythagorean discoveries along with later work, culminating in the construction of the five "Platonic solids," the only solid figures that are regular (having identical equal-sided faces): the tetrahedron, the cube, the octahedron, the dodecahedron, and the icosahedron (figure 1.1). The major contribution of the Pythagoreans, though, was the concept of mathematical proof, the idea that one could construct irrefutable demonstrations of theoretical propositions that would admit of no

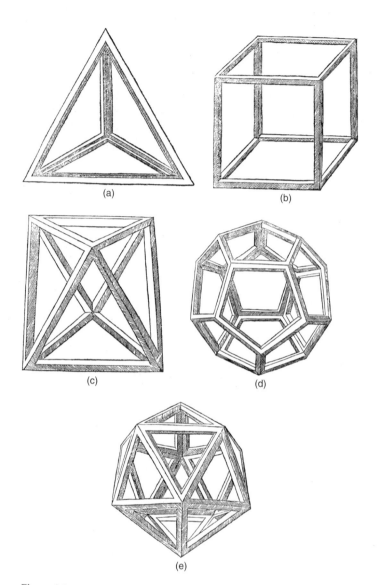

(a)

(b)

(c)

(d)

(e)

**Figure 1.1**
The five regular Platonic solids, as illustrated after Leonardo da Vinci
in Luca Pacioli, *On the Divine Proportion* (1509). a. tetrahedron, b. cube,
c. octahedron, d. dodecahedron, e. icosahedron.

exception. Here they went beyond the Babylonians, who, despite their many accomplishments, seem not to have been interested in proving propositions. The Pythagoreans and their followers created "mathematics" in the sense we still know it, a word whose meaning is "the things that are learned," implying certain and secure knowledge.

The myths surrounding the Pythagorean brotherhood hide exactly who made their discoveries and how. Pythagoras himself was said to have recognized the proportions of simple whole-number ratios in the musical intervals he heard resounding from the anvils of a blacksmith shop: the octave (corresponding to the ratio 2:1), the fifth (3:2), the fourth (4:3), as ratios of the weights of the blacksmith's hammers. This revealed to him that music was number made audible. (This is a good point to note an important distinction: the modern fraction $\frac{3}{2}$ denotes a breaking of the unit into parts, whereas the ancient Greeks used the ratio 3:2 to denote a relation between unbroken wholes.) Another story tells how he sacrificed a hundred oxen after discovering what we now call the Pythagorean Theorem. These stories describe events that were felt to be of such primal importance that they demanded mythic retelling.

There is a third Pythagorean myth that tells of an unforeseen catastrophe. Despite their motto that "all is number," the Pythagoreans discovered the existence of magnitudes that are radically different from ordinary numbers. For instance, consider a square with unit side. Its diagonal cannot be expressed as any integral multiple of its side, nor as any whole-number ratio based on it. That is, they are *incommensurable*. Box 1.1 describes the simple argument recounted by Aristotle to prove this. This argument is an example of a *reductio ad absurdum*: We begin by assuming hypothetically that such a ratio exists and then show that this assumption

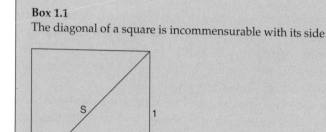

**Box 1.1**
The diagonal of a square is incommensurable with its side

Let the square have unit side and a diagonal length $s$. Then suppose that $s$ can be expressed as a ratio of two whole numbers, $s = m{:}n$. We can assume further that $m$ and $n$ are expressed in lowest terms, that is, they have no common factors. Now note that $s^2 = m^2{:}n^2 = 2{:}1$, since the square on the hypotenuse $s$ is double the square on the side, by the Pythagorean theorem. Therefore $m^2$ is even (being two times an integer), and so too is $m$ (only the square of an even number is even). But then $n$ must be odd, since otherwise one could divide $m$ and $n$ by a factor of 2 and simplify them further. If $m$ is even, we can let $m = 2p$, where $p$ is some number. Then $m^2 = 4p^2 = 2n^2$, and $n^2 = 2p^2$. But this means that $n^2$ is even, and so too is $n$. Since a whole number cannot be both even and odd, our original assumption that $s = m{:}n$ must be wrong. Therefore the diagonal of a square cannot be expressed as a ratio of two whole numbers.

leads to an absurdity, namely that one and the same number must be both even and odd. Thus the hypothesis must have been wrong: no ratio can represent the relation of diagonal to side, which is therefore *irrational*, to use a modern term.

The original Greek term is more pungent. The word for ratio is *logos*, which means "word, reckoning, account," from a root meaning "picking up or gathering." The new magnitudes are called *alogon*, meaning "inexpressible, unsayable." Irrational magnitudes are logical consequences of geometry, but they are inexpressible in terms of ordinary numbers, and the Greeks were careful to use entirely different words to denote a number (in Greek, *arithmos*) as opposed to a magnitude (*megethos*). This distinction later became blurred, but for now it is crucial to insist on it. The word *arithmos* denotes the counting numbers beginning with two, for "the unit" or "the One" (the Greeks called it the "monad") was not a number in their judgment. The Greeks did not know the Hindu-Arabic zero and surely would not have recognized it as an *arithmos*; even now, we do not hear people counting objects as "zero, one, two, three, . . . ." Thus, the expression "there are no apples here" means "there aren't any apples here" more than "there are zero apples here."

It was only in the seventeenth century that the word "number" was extended to include not only the counting numbers from two on, but also irrational quantities. Ancient mathematicians emphasized the distinctions between different sorts of mathematical quantities. The word *arithmos* probably goes back to the Indo-European root *(a)rī*, recognizable in such English words as *rite* and *rhythm*. In Vedic India, *ṛta* meant the cosmic order, the regular course of days and seasons, whose opposite (*anṛta*) stood for untruth and sin. Thus, the Greek word for "counting number" goes back to a concept of cosmic order, mirrored in proper ritual: certain things must come *first*, others *second*, and so on. Here, due order is important; it is not possible to stick in upstart quantities like "one-half" or (worse still) "the square root of 2" between the *integers*, a word whose Latin root means unbroken or whole.

The integers are paragons of integrity; they should not be confused with magnitudes, which are divisible.

At first, the Pythagoreans supposed that all things were made of counting numbers. In the beginning, the primal One overflowed into the Two, then the Three, then the Four. The Pythagoreans considered these numbers holy, for $1 + 2 + 3 + 4 = 10$, a complete decade. They also observed that the musical consonances have ratios involving only the numbers up to four, which they called the "holy Tetractys." Out of such simple ratios, they conjectured, all the world was made. The discovery of magnitudes that cannot be expressed as whole-number ratios was therefore deeply disturbing, for it threatened the entire project of explaining nature in terms of number alone. This discovery was the darkest secret of the Pythagoreans, its disclosure their greatest scandal. The identity of the discoverer is lost, as is that of the one who disclosed it to the profane world. Some suspect them to have been one and the same person, perhaps Hippias of Mesopontum, somewhere around the end of the fifth century B.C., but probably not Pythagoras himself or his early followers. Where Pythagoras had called for animal sacrifice to celebrate his theorem, legend has it that the irrational called for a human sacrifice: the betrayer of the secret drowned at sea. Centuries later, the Alexandrian mathematician Pappus speculated that

they intended by this, by way of a parable, first that everything in the world that is surd, or irrational, or inconceivable be veiled. Second, any soul who by error or heedlessness discovers or reveals anything of this nature in it or in this world wanders on the sea of nonidentity, immersed in the flow of becoming, in which there is no standard of regularity.

Those who immerse themselves in the irrational drown not by divine vengeance or by the hand of an outraged brotherhood but in the dark ocean of nameless magnitudes. Ironically, this is a consequence of geometry and the Pythagorean Theorem itself. When Pythagoras realized that the square on the hypotenuse was equal to the sum of the squares on the other two sides, he was very close to the further realization that, though the squares might be commensurable, the sides are not. Indeed, the argument given in box 1.1 depends crucially on the Pythagorean Theorem. That argument suggests that, had Pythagoras tried to express the ratio of the diagonal of a square to its side, he would have realized its impossibility immediately. He probably did not take this step, but his successors did.

The discovery of the irrational had profound implications. From it, Pappus drew a distinction between such "continuous quantities" and integers, which "progressing by degrees, advance by addition from that which is a minimum, and proceed indefinitely, whereas the continuous quantities begin with a definite whole and are divisible indefinitely." Further, if we start with an irreducible ratio such as 2:3, we can build a series of similar ratios in a straightforward manner: $2:3 = 4:6 = 6:9 = \cdots$. But if there is no smallest ratio in a series, there can be no ratio expressing the whole. Pappus's words suggest that it was this argument that may have opened the eyes of the Pythagoreans. Consider again the diagonal and side of a square. The attempt to express both of them as multiples of a common unit requires an infinite regress (box 1.2). However small we take the unit, the argument requires it to be smaller still. Again we see that no such unit can exist.

The challenge of Greek mathematics was to cope with two incommensurable mathematical worlds, arithmetic and geometry, each a perfect realm of intelligible order within itself, but with a certain tension between them. In Plato's dialogues,

**Box 1.2**
A geometric proof of the incommensurability of the diagonal of a square to its side, using an infinite regress:

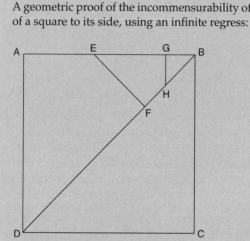

In the square *ABCD*, use a compass to lay off *DF* = *DA* along the diagonal *BD*. At *F*, erect the perpendicular *EF*. Then the ratio of *BE* to *BF* (hypotenuse to side) will be the same as the ratio of *DB* to *DA*, since the triangles *BAD* and *EFB* are similar. Suppose that *AB* and *BD* were commensurable. Then there would be a segment *I* such that both *AB* and *BD* were integral multiples of *I*. Since *DF* = *DA*, then *BF* = *BD* − *DF* is also a multiple of *I*. Note also that *BF* = *EF*, because the sides of triangle *EFB* correspond to the equal sides of triangle *BAD*. Further, *EF* = *AE* because (connecting *D* and *E*) triangles *EAD* and *EFD* are congruent. Thus, *AE* = *BF* is a multiple of *I*. Then *BE* = *BA* − *AE* is also a multiple of *I*. Therefore, both the side (*BF*) and hypotenuse (*BE*) are multiples of *I*, which therefore is a common measure for the diagonal and side of the square of side *BF*. The process can now be repeated: on *EB* lay off *EG* = *EF* and construct *GH* perpendicular to *BG*. The ratio of hypotenuse to side will still be the same as it was before and hence the side of the square on *BG* and its diagonal also share *I* as a common measure. Because we can keep repeating this process, we will eventually reach a square whose side is less than *I*, contradicting our initial assumption. Therefore, there is no such common measure *I*.

this challenge elicits deep responses, reaching out from mathematics to touch emotional and political life. A pivotal moment occurs in the dialogue between Socrates and Meno, a visiting Thessalian magnate who was a friend and ally of the Persian king. Meno was notoriously amoral, a greedy and cynical opportunist. Strangely, on his last day in Athens he asks Socrates over and over again whether virtue can be taught or comes naturally. Their conversation turns on the difference between true knowledge and opinion.

At the heart of their discussion, Socrates calls for a slave boy, with whom he converses about how to double the area of a square of a given size. Unlike Meno, the boy is innocent and frank; he confidently expresses his opinion that if you double the side of a square, you double its area. Their conversation is a perfect example of Socrates' practice of philosophy through dialogue. As they talk, the boy realizes that a square of double side has *four* times the area, which leaves him surprised and perplexed. The Greek word for his situation is *aporia*, which means an impasse, an internal contradiction. Just before this conversation, Socrates' questioning had revealed contradictions in Meno's confident opinions about virtue, and Meno had lashed out angrily. Socrates, he said, was like an ugly stingray that harms his victims and renders them helpless. Socrates' answer is to show how well the slave boy could take being "stung." The boy is amazed and curious, not angry. He readily follows Socrates' lead in drawing a new picture (box 1.3). In a few strokes, the real doubled square emerges by drawing the diagonals within the boy's fourfold square. Responding freely to Socrates' suggestions, the boy grasps this himself. Meno is forced to admit that the "sting" of realizing his ignorance did not harm the boy, who replaced his false opinion with a true one. The dialogue ends with Meno smoldering, foreshadowing the angry Athenians who

**Box 1.3**
Socrates' construction of the doubled square in the *Meno*

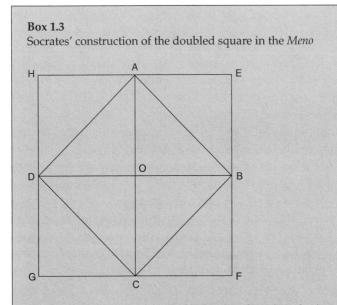

Let the original square be *AEBO*. The slave boy thought that the square on the doubled side *HE* would have twice the area, but realized that in fact that *HEFG* has four times the area of *AEBO*. At Socrates' prompting, he then draws the diagonals *AB, BC, CD, DA* within the square *HEFG*. Each triangle *AOB = BOC = COD = DOA* is exactly half the area of the original square, so all four of these together give the true doubled square *ABCD*.

later voted to execute the philosopher. Their outrage points to the power of the new mathematical insights. Though Socrates had not referred to the irrationality of the diagonal, it was crucial. The process of doubling a square (an eminently rational enterprise) required recourse to the irrational, a fact that was not lost on Plato or his hearers.

Though a consequence of logical mathematics, clearly the word "irrational" already had acquired the emotional connotations it still retains. In Plato's *Republic*, Socrates jokes that young people "are as irrational as lines" and hence not yet suited to "rule in the city and be the sovereigns of the greatest things." Appropriately and yet ironically, Socrates prescribes mathematics, along with music and gymnastics, for these young irrationals to tame what is most disorderly and incommensurate in their souls. His joke points to a widely held sense that irrationality in mathematics was a troubling sign of confusion and disorder in the world, a danger as fearful as drowning. Certainly the Pythagoreans took this dire view, but Plato's dialogues open a larger perspective. What is irrational, in the soul or in mathematics, may be harmonized with the rational; to use an unforgettable image from another dialogue, the black horse of passion may be yoked to the white horse of reason.

Plato's great dialogue on the nature of knowledge rests on this mathematical crux. It is named after Theaetetus, a mathematician who is introduced to us as he is being carried back to Athens, dying from battlefield wounds and dysentery. In a flashback to Theaetetus' youth, we learn that he made profound discoveries about the irrationals and the five regular solids, and he conversed with Socrates shortly before the philosopher's trial and death. Socrates was deeply impressed with this youth, who seemed destined to do great things and who also resembled Socrates physically, down to the snub nose and bulging eyes. Also present during their conversation was Theaetetus' teacher Theodorus, an older mathematician who had proved the irrationality of $\sqrt{3}$, $\sqrt{5}$, $\sqrt{7}$, . . ., all the way to $\sqrt{17}$, where for some reason he stopped.

Socrates' characteristic irony is not in evidence as he questions Theaetetus, who explains his discovery that there are degrees of irrationality. Though such magnitudes as the square roots of 3 or of 17 are irrational, they are still "commensurable in square," since their squares have a common measure (that is, since $\left(\sqrt{3}\right)^2 = 3$ and $\left(\sqrt{17}\right)^2 = 17$ are both integers). Socrates is struck with the truth and beauty of these insights and uses them as examples that lead to a broader conversation about the nature of knowledge. He reminds Theaetetus and Theodorus that he has a reputation as one who "stings" by inducing perplexity and jokingly asks Theaetetus not to denounce him as an evil wizard, explaining that he is really a midwife who helps people deliver themselves of their conceptions.

Anticipating his indictment on the very next day, Socrates justifies himself not to his angry accusers but to this gentle and gifted young man, so much like himself. Far from feeling antagonistic, Theaetetus is ready to enter a searching inquiry that begins with mathematics as a touchstone of true knowledge, testing whether other knowledge comes through the senses or more mysteriously from within the soul. Though he depicts himself as sterile, barren of wisdom, Socrates helps Theaetetus bring his conception to birth and tests its health. Socrates had often made fun of his own ugly features, but he describes Theaetetus as beautiful. Theaetetus' mathematical insight is commensurate with the bravery that will allow him to fight for his city and die with exceptional honor. Such is the courage of one who could wrestle with the irrational.

During their conversation, Socrates encourages his guests to "put themselves to torture," by which he means that they should struggle fearlessly to test and refine their opinions together. In Greek, the word for "torture" can also mean the "touchstone," a mineral that is able to distinguish gold from base metal by the mark each makes on it. This extreme metaphor has overtones of the judicial torture used to coerce truth from slaves, but Socrates uses it to signify a search for truth that defies even intense pain and humiliation. Like soldiers or athletes, Socrates and Theaetetus see in suffering the path to the superlative pleasure of ultimate truth. This they have learned from mathematics, whose study often seems painful to those who do not know the pleasure of insight. It is no wonder that Plato placed over the door to his academy the admonition: "Let no one ignorant of geometry enter here."

The discoveries of Theaetetus and the test of mathematical proof were enshrined in Euclid's *Elements*, which remains even today a living fountainhead of mathematics, invaluable for beginners as well as experienced mathematicians. Beyond presenting his own results, Euclid set the discoveries of others in order as a touchstone of mathematical lucidity and logical force. In the case of the irrational, Euclid drew on a compromise introduced by Eudoxus, who kept numbers and irrational magnitudes strictly apart, but yet in proportion. For instance, Euclid considers two numbers in a certain ratio (say 2:3) and shows how this proportion could be equal to that between two irrational magnitudes (as $2\sqrt{2}:3\sqrt{2}$ is equal to 2:3). Nevertheless, he would never mix the two distinct types so as to allow a ratio between a number and a magnitude. This was not simply mathematical apartheid but a decision to consider numbers and magnitudes as entirely distinct genera, whose mixing might lead to incalculable confusion.

Euclid's contribution went far beyond this separation of realms. In Book V, he introduces a far-reaching definition of equality or inequality that extends to ratios of irrational magnitudes. Following Eudoxus, he proposes that if we want to check whether two ratios are equal, we should multiply the terms by various integers to check whether these multiples are respectively greater, equal, or less (box 1.4). This

> **Box 1.4**
> Euclid's definition of equal ratio, which is applicable to any
> magnitude (Book V, Definition 5)
>
> The ratio $a:b$ is said to be equal to the ratio $c:d$ if, for any
> whole numbers $m$ and $n$, when $ma$ is compared with $nb$ and
> $mc$ is compared with $nd$, the following holds: if $ma > nb$,
> then $mc > nd$; if $ma = nb$, then $mc = nd$; and if $ma < nb$, then
> $mc < nd$.

definition of equality still depends on testing multiples of
magnitudes, even though the magnitudes themselves may
have no common measure. It also uses *any* multiples whatever, as if to examine all possible multiples in order to determine whether the multiplied ratios could ever be equal.
Thus, it is really a *test*, a trial by multiplication, a way to nav-
igate the irrational sea. Euclid puts it in strong contrast with
the way he treats whole numbers in Books VI and VII, for
integers are commensurable because they have a common
unit.

Euclid's most daring inquiry into the irrational occurs in
Book X, which asks: Do the irrational magnitudes have some
intelligible order? Can one classify them into clear categories
by genus and species? He begins by showing that any magnitude can be indefinitely divided. Though implicit in geometry, he brings into prominence what later came to be called the
*continuum*, meaning a continuously and endlessly divisible
magnitude, as opposed to the indivisible One, whose integral multiples constitute all the counting numbers. To show
this indefinite divisibility, Euclid demonstrates how we can
successively subtract from any magnitude half or more of
that magnitude, and then keep repeating this process until

**Box 1.5**
Euclid's statement of the indefinite divisibility of any magnitude (Book X, Proposition 1)

Take half (or more) of the given magnitude, and then the same proportion of what remains, and the same proportion yet again of what remains, continuing the process as far as necessary so that the remainder can be made less than any given line.

finally we have left a magnitude that is smaller than any given amount (box 1.5). Thus, there is no smallest magnitude, no geometrical "atom" or least possible magnitude making up all others, for if there were, all magnitudes would share that smallest magnitude as their common measure. Here again, Euclid sets in play a process indefinitely repeated, not picturable in a single figure but intelligible and logically compelling, nonetheless.

Then Euclid sets out to classify different kinds of irrationals, naming them and showing their interrelations. As Theaetetus had shown, irrationality is a relative term. The diagonal of a square is irrational compared to the side, but it can be commensurable with another line, which might be the side or diagonal of another square. What is speakable depends above all on the *relation* between figures. Euclid's classification of irrationals is intricate, though it does not go beyond what we would call the square root of the sum or difference of two square roots. He identifies such quantities in the division of a geometrical line, but we can also divide a string to make it sound different intervals. This means we can formulate a musical version of the mathematical crisis of the irrational. If we try to divide an octave (whose ratio

**Box 1.6**
The sound of square roots

Take two strings, one sounding an octave higher than the other, so that their lengths are in the ratio 2:1. Then find the geometric ratio (also called the mean proportional) between these strings, the length $x$ at which 2:$x$ is the same proportion as $x$:1. This means that 2:$x$ = $x$:1; cross-multiplying this gives $x^2 = 2$. Thus, the "ratio" needed is $\sqrt{2}$:1 $\approx$ 1.414, in modern decimals. This is close to the dissonant interval called the tritone, which later was called the "devil in music," namely the interval composed of three equal whole steps each of ratio 9:8. The tritone is thus 9:8 × 9:8 × 9:8 = $9^3$:$8^3$ = 729:512 $\approx$ 1.424.

is 2:1) exactly at the point of the geometric mean, we get the mongrel "ratio" $\sqrt{2}$:1 (box 1.6). This is very close to the highly dissonant interval later called the "devil in music," the tritone. If the whole universe is based on number, such harmonic problems are critical.

Euclid presents his classification of irrationals through a hundred careful propositions. After this tour de force, he says something amazing in the final proposition of the book: From the lines already drawn, one can go on to define still other irrational lines that are "infinite in number, and none of them is the same as any of the preceding." Although his tone is impassive, this is a portentous statement. The realm of the irrational is infinite not just because there are an unlimited number of irrational magnitudes of each type but even more because there is an infinite variety of *kinds* of such magnitudes, each a different species with infinitely many examples. The discovery of the irrational disclosed an infinitely branching path.

Euclid's impassive tone does not disclose what he thought of this situation. By this final proposition, Euclid could have meant to indicate a disturbing glance into the irrational abyss, as if to say: Here lies an unfathomable, trackless sea of endlessly different magnitudes, from which one should turn away in horror. But there is another possible reading of his silence. He might have meant: Here lies an inexhaustible store of treasures, infinite in number though each is finite in magnitude. Behold, and wonder.

# 2      Controversy and Coefficients

In his attempt to survey the irrational magnitudes, Euclid was laying out possible solutions of what we now call the quadratic equation. But that statement distorts his own way of understanding these matters and thus is seriously misleading. The ancient Greek mathematicians did not know the word "algebra," or the symbolic system we mean by it, even though many of their propositions can be translated into algebraic language. Euclid kept numbers rigorously separate from lines and magnitudes. He might have been horrified by the way algebra mixes rational and irrational, numbers and roots. So, as we prepare to consider the meaning of equations, we need to pause and assess the significance of the mathematical revolution reflected in the rise of algebra as we know it.

The word "algebra" is Arabic, not Greek, and refers to the joining together of what is broken. Thus, in *Don Quixote*, Cervantes calls someone who sets broken bones an *algebrista*. The original Arabic algebraists created recipes for solving different kinds of problems, usually involving whole numbers. They worked by presenting examples with solutions, but without the methodical symbolism we now associate with algebra and without the systematic processes of solution that are equally important. They did recognize the logical force of geometry and tried to illustrate their solutions with geometric examples, but they could not go far without the symbols that might have enabled them to speak with more generality.

What is surprising, then, is that some of the greatest advances in solving algebraic problems came long before such powerful symbols were devised. In the case of the quadratic equation, the Babylonians were able to solve many similar problems four thousand years ago. Notably more advanced than the Egyptians, the Babylonians made difficult calculations using a notation essentially equivalent to modern place-value, but based on the number sixty (box 2.1). We still use sexagesimal notation in our measurements of time (60 minutes in the hour) and of angle (60 minutes of arc in a degree). Such a notation of place-value was unknown in Europe until the Renaissance, and the rich achievements of Babylonian mathematics came fully to light only in the twentieth century. When they considered a problem, the Babylonians wrote it out and solved it in words. With no symbolic notation for unknowns, they could not express the *general* solution of an equation with arbitrary values, though they could solve specific examples, as in box 2.2.

Similar statements could be made about the methods of the Arabic algebraists, whose principal representative to the

---

**Box 2.1**
Babylonian sexagesimal (base 60) notation

The Babylonian notation for our decimal number 870 can be written as $14, 30 = 14(60)^1 + 30(60)^0$; the notation for the modern fraction 0.5 would be $0; 30 = \frac{30}{60} = 0(60)^0 + 30(60)^{-1}$.

> **Box 2.2**
> Babylonian solution
>
> Here is an example of a Babylonian solution of what we would call the equation $x^2 - x = 870$, with the numerical values translated into decimal notation: "Take half of 1, which is 0.5, and multiply 0.5 by 0.5, which is 0.25; add this to 870 to get 870.25. This is the square of 29.5. Now add 0.5 to 29.5, and the result is 30, the side of the square." Compare this with the modern solution of the equation $a_2 x^2 + a_1 x + a_0 = 0$, with $a_2 = 1$, $a_1 = -1$, and $a_0 = -870$, namely $x = \left(-a_1 \pm \sqrt{a_1^2 - 4a_2 a_0}\right) / 2a_2$ gives $x = \sqrt{\left(\frac{1}{2}\right)^2 + 870} + \frac{1}{2}$, each term of which corresponds to a step in the Babylonian solution.

West was Muhammed ibn-Musa al-Khwārizmī. Writing in the ninth century A.D., he sets forth *al-jabr* as an eminently practical art, "such as men constantly require ... in all their dealings with one another"; he gives as instances the problems that emerge from inheritances, lawsuits, trade, land measurement, and canal building. Like the Babylonians, al-Khwārizmī focuses on specific problems, but he adds that "it is necessary that we should demonstrate geometrically the truth of the same problems which we have explained in numbers." Box 2.3 shows his geometrical explanation of one problem: to solve for the unknown "root" if "a square and 10 roots equal 39 units" ($x^2 + 10x = 39$, as we would say). His method is that of "completing the square," though al-Khwārizmī confines it to this one example. Compare the general expression of this method in modern symbols shown in box 2.3. Shortly, we will consider the history and significance of these symbols.

**Box 2.3**

Al-Khwārizmī's geometric explanation of the solution to quadratic equations

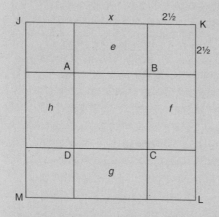

Consider the proposition that "a square and 10 roots equal 39 units" ($x^2 + 10x = 39$, in modern notation). Let the central square $ABCD$ have area $x^2$; its side is thus $x$. Add to this square four equal areas, labeled $e$, $f$, $g$, $h$, each of area $2\frac{1}{2}x$, so that together they have area $10x$. Each of these is a rectangle of sides $x$ and $2\frac{1}{2}$. Now complete the larger square $JKLM$ by including the little squares at each corner, whose area is each $2\frac{1}{2} \times 2\frac{1}{2} = 6\frac{1}{4}$; the total area of all four corner squares is $4 \times 6\frac{1}{4} = 25$. We are given that the central square plus the rectangles equals 39, so the large square $JKLM$ has area $39 + 25 = 64$ and hence side 8. But the side of the large square is $2\frac{1}{2} + x + 2\frac{1}{2} = 5 + x = 8$, so we conclude that $x = 3$.

In comparison, the algebraic solution of $a_2x^2 + a_1x + a_0 = 0$ is $x = \left(-a_1 \pm \sqrt{a_1^2 - 4a_2a_0}\right)/2a_2$. In this case, $a_2 = 1$, $a_1 = 10$, $a_0 = -39$, so $x = (-10 \pm \sqrt{100 + 156})/2 = (-10 \pm 16)/2 = 3$ or $-13$.

As the algebraic writings of the Arabs were translated into Latin, Europeans began to work in what they called "the great art," also called the art of the coss or *cosa*, meaning the "thing" or unknown. The story of early modern algebra is tumultuous; after long periods of inactivity, great strides were made in a matter of years by a few individuals. These mathematical innovations were often related to problems raised by economic changes, as the Arabs' work had also been. During the centuries following the Crusades, new conditions of commerce involving extensive traffic in commodities required widespread use of currency as a medium of exchange and a new way of analyzing the flow of assets. European goods ranged widely, traded for spices and silks. As merchants strove for rapid increase of capital, they found they needed new ways to deal with dispersed partners and factors operating in diverse local economic situations. Marine insurance emerged, as did bills of exchange in draft form and modes of international credit. The ancient methods of bookkeeping also broke down when, as Shylock describes it, a trader has "an argosy bound to Tripolis, another to the Indies; I understand, moreover, upon the Rialto, he hath a third at Mexico, a fourth for England—and other ventures he hath, squand'red abroad."

In this new situation, double-entry bookkeeping was an invaluable innovation. The basis of all subsequent accounting practice, this system requires a twofold listing of all debits and credits, recorded in a single currency and checked by balancing both the debit and the credit sides of the ledger. The new bookkeeping was extremely awkward with Greek or Roman numerals but flourished with the advent of Hindu-Arabic numerals, which Leonardo of Pisa (better known as Fibonacci) helped introduce to Europe in his *Book of the Abacus* (1202). At this time, the Arabic word *ṣifr* gave birth to the English words "zero" and "cipher," essential concepts for reckoning with numbers as a symbolic code.

Fibonacci's book concerns problems in commercial transactions, especially currency exchanges that involve fractions. Such practical economic problems lead to more speculative questions: "How many pairs of rabbits will be produced in a year, beginning with a single pair, if in every month each pair bears a new pair that becomes productive from the second month on?" Considering such problems leads Fibonacci to his celebrated sequence: 1, 1, 2, 3, 5, 8, 13, ..., in which each term is the sum of the two immediately preceding it. This sequence could describe the growth of capital as well as an increasing population of rabbits. Later, the quantity now called $e = 2.718...$, the key to exponential growth, emerges during a further consideration of compound interest. Elsewhere, Fibonacci also considers cubic equations, which he argues cannot be solved with integers, square roots, and fractions, but for which he presents quite accurate approximations.

Using Fibonacci as its main source, Luca Pacioli's *Summary of Arithmetic, Geometry, Proportions, and Proportionality* (1494) includes the first treatise on double-entry bookkeeping. Pacioli was a friend and colleague of the artists Piero della Francesca (from whom he plagiarized his mathematics) and Leonardo da Vinci, who illustrated his book *On the Divine Proportion* (1509), a treatment of the "golden ratio" of the Greeks. Pacioli considers that it is necessary in business to be "a good accountant and a ready mathematician" and that business affairs must be arranged systematically, "for without systematic recording, their minds would always be so tired and troubled that it would be impossible for them to conduct business." The system he expounds has the great advantage that it envisions an exhaustive accounting in which all the enterprises of a company can be drawn together.

It is appropriate that Pacioli's *Summary* was one of the first books to be printed on movable type, since it appealed to the increasing audience of businessmen who needed systematic and ironclad schemes of accounting for their businesses. Perhaps the best known aspect of the double-entry system is the duality between credit and debit, manifest in the dual recording of every transaction on both sides of the ledger; this leads to the possibility of double-checking the accuracy of accounts through the comparison of "trial balances" struck independently on each side of the ledger. Though this is surely useful, it is not the crucial innovation of the double-entry scheme, for the accounts of Roman slaves or medieval factors were also able to be balanced. Double entry allows not only balance but *completeness* of accounting because it records *every* kind of transaction, including both "real accounts" that represent concrete transactions with a specific client or creditor and what now are called "nominal accounts," such as "depreciation" or "good-will."

Pacioli places his discussion of bookkeeping and commercial mathematics next to his exposition of algebra. Here he was not an innovator but an influential compiler of techniques. His exposition indicates the close association between commercial and what we would consider "pure" mathematics. We gain a similar impression from other early works, such as the "Treviso Arithmetic" (1478) and Johann Widman's *Mercantile Arithmetic* (1489), the oldest book in which the familiar "+" and "−" signs appear in print. Here again these symbols at first refer to surplus and deficiency in warehouse inventory, only later becoming signs of abstract operations. As with Pacioli, practical considerations lead surprisingly quickly to questions that transcend the symbols' commercial origins, including questions about the solvability of equations.

Though the Babylonians had made significant strides in solving what we would call cubic equations, all of their work had been lost. Fibonacci thought such equations could not be solved by magnitudes and square roots; Pacioli thought them quite unsolvable by algebra, an opinion he derived from the Arabic poet-mathematician Omar Khayyām, who had done extensive work classifying cubics and exploring their solvability. Piero della Francesca, a great mathematician as well as painter, tried unsuccessfully to solve cubic, quartic, and even quintic equations. The breakthrough came from a motley bunch of Italians, of whom the most colorful was Girolamo Cardano, whose life spanned most of the sixteenth century (figure 2.1). Cardano was a physician, an inveterate gambler, and an astrologer, whose religious opinions were sufficiently heretical that the Inquisition jailed him in his old age. It was his interest in gambling that led Cardano to write the first book on the mathematics of probability. Despite all this, he held prestigious professorships and eventually received a pension from the pope. He wrote prolifically, and his *Book of My Life* is astonishingly frank, detailing not only his own passions and obsessions but also the misadventures of his children (one son poisoned his wife and was executed, the other was a ne'er-do-well who robbed his father). His role in the history of algebra is central but controversial, the subject of contention even among his contemporaries.

In his book *The Great Art, or the Rules of Algebra* (1545), Cardano treats problems that we would call quadratic as already solved, and he cites "Mahomet the son of Moses the Arab" (that is, al-Khwārizmī) as well as Fibonacci. Cardano gives the problem of the cubic equation a rhetorical form and a commercial context: "Two men go into business together and have an unknown capital. Their gain is equal to the cube of the tenth part of their capital. If they had made

**Figure 2.1**
Girolamo Cardano.

three ducats less, they would have gained an amount exactly equal to their capital. What was the capital and their profit?" In modern terms, if the unknown capital is $x$, then the gain is $\left(\frac{x}{10}\right)^3$ and the equation defined by the problem is $\left(\frac{x}{10}\right)^3 - 3 = x$.

The commercial aspect of such problems may have been less important to Cardano than the possibility of rivaling the ancients. In those days, classical statues were regularly being unearthed in Italy, astonishing beholders with the immense achievements of antiquity. The works of Plato had only recently been translated into Latin, and the Greek mathematicians were likewise just beginning to be read. Were these ancients gods, not to be touched by the moderns? Yet the ancients had not solved the cubic, and here the moderns might still emulate and even surpass them.

The chain of events is not completely clear; what is clear is that the solution of the cubic became the first example of a sordid modern genre: the scientific priority fight. Niccolò Fontana, nicknamed Tartaglia ("Stammerer"), claimed to have solved the problem, but in truth he started his work after had heard that it had already been solved by Scipione del Ferro, who did not publish it but disclosed it to a student before his own death. Tartaglia may well have had a hint from this earlier source. In those days, standards of attribution and citation were lax; Tartaglia had already published someone else's translation of Archimedes as if it were his own, and he also tried to pass as the author of the law of the inclined plane without giving proper credit to its true discoverer. At any rate, Tartaglia did make a breakthrough in solving cubic equations, which allowed him to win a public contest with del Ferro's student. These contests show something of the status of algebra at the time; they were rowdy tournaments of skill, in which each contestant would try to stump the other with a problem to win a handsome prize. The popular appeal of these mathematical gladiatorial matches was on a par with bear-baiting and involved something of the same atmosphere.

But how could Tartaglia win, if del Ferro's student had learned the solution from his teacher? In those days, there were many variations of cubic problems that seemed separate, and to which various techniques were applied, because mathematicians lacked the modern symbolism that can include all possible varieties of cubic equations under the general form $a_3x^3 + a_2x^2 + a_1x + a_0 = 0$. Thus, the student might have been able to solve a problem like "the cube and the number equal to the first power," as Cardano calls it (for instance, $x^3 + 21 = 16x$), but still be stymied by "the cube and first power equal to square and number" (as he would have written an equation like $x^3 + 20x = 6x^2 + 33$). On a more basic level, since only positive integers were then recognized, any case involving negative numbers would have had to be treated separately, by moving such numbers to the other side of the equation. In any case, news of Tartaglia's victory reached Cardano, who pressed him for details of his methods, hinting that he would arrange a meeting with a wealthy patron.

Tartaglia was wary but needed money, perhaps hindered in earning his living because of a speech impediment caused by saber wounds near the mouth that he received as a child during the French invasion of Italy in 1512. He disclosed his secret to Cardano in 1539, but only after making him swear not to reveal it. Tartaglia communicated the solution in the form of a cumbersome poem, almost riddling enough to be in code. Years passed, during which Tartaglia did not publish his solution. Then, in 1543, Cardano visited Bologna and gained access to del Ferro's papers, which showed that del Ferro had in fact solved the cubic before Tartaglia. Cardano then felt free to publish the method in his *Great Art*, attributing it to Tartaglia, and del Ferro before him, but without mention of his oath of secrecy. By giving credit, he surely sought to avoid the appearance of trying to steal the secret. He says that before seeing this work, he had believed the claim of

Luca Pacioli that there was no rule that solved equations beyond the quadratic. But by mentioning del Ferro's earlier discoveries, he diminishes Tartaglia's claim to uniqueness. In explanation, he only remarks that, in response to his entreaties, Tartaglia "gave" it to him and also that, after learning it, he went further than Tartaglia ever did.

In fact, Tartaglia did not have the complete solution to every variety of cubic that Cardano presents, which he divides into thirteen cases, many with subcases. Still, Tartaglia was outraged and published accusations of plagiarism. At this point a new challenger stepped in. Cardano had a servant named Ludovico Ferrari who showed such keen interest in mathematics that Cardano took him on as a student. With his master's honor at stake, Ferrari challenged Tartaglia to a mathematical duel that ended with Ferrari's victory, or so it seems from subsequent events, for Ferrari obtained many flattering offers, including a professorship in Bologna, while Tartaglia's account reveals that he left even before the contest was over.

How, then, was the cubic equation solved? The method turns out to be an early instance of a powerful mathematical strategy: to solve a more difficult problem, reduce it to a simpler problem you have already solved. Here, the technique of "completing the square" is the key. As noted earlier, we can solve a quadratic equation by reducing it to the form of a perfect square and then taking the square root. It is natural to try to attack the cubic in the same way by "completing the cube." Del Ferro, Tartaglia, and Cardano showed how any given cubic equation can be put in the form of a perfect cube equal to some numerical value. Then we can complete the solution by taking the cube root. Though they did not present their deduction with a three-dimensional diagram, it is useful to consider such a picture (box 2.4), for (like the parallel picture shown for the quadratic equation in box 2.3) it shows that this problem is essentially like a jigsaw puzzle. The solution of this three-dimensional puzzle requires breaking it into slices.

That is, completing the cube involves a series of interlocking solids that must be completed and reassembled. Box 2.4 shows an example from Cardano, along with its solution in modern notation. Though the result looks a little daunting, it only involves nested radicals, that is, roots of roots (such as the square root of a cube root). Just as the solution of the quadratic involves square roots, the solution of the cubic involves cube roots, which in turn contain expressions involving square roots. Later, we will see that this characteristic structure of nested radicals for the solution of the cubic gives an important clue in the search for the solution of higher-order equations beyond the cubic.

Ferrari not only defended his master but also made an advance of his own. At Cardano's request, Ferrari addressed the problem of solving what we would call equations of the fourth degree, or quartics, which they called "square-square." Their use of this term shows that they did not yet have our more general conception of equations of arbitrary degrees. In Cardano's book, cubic problems seem entirely separate from quadratic ones, and both have a geometric significance that is lacking in the case of "square-square" equations. Because cubic equations can be solved by a three-dimensional puzzle, one might think that quartic equations would somehow require struggling with a puzzle in four dimensions. In fact, the solution of the quartic does not require a fourth spatial dimension. It turns out only to require completing the square in a somewhat different way, using $x^2$ as the

**Box 2.4**
Cardano's method of completing the cube

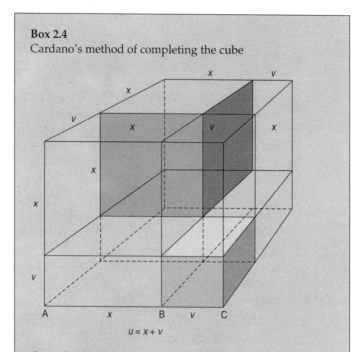

$u = x + v$

Consider Cardano's example "Let the cube and six times the side be equal to 20," or $x^3 + 6x = 20$, in modern notation. We want to make this problem take the form of a perfect cube, and the trick needed is to set $x = u - v$. In the diagram, let $x^3$ be the upper shaded cube, now embedded in a larger cube with side $AC = u$. There is also a lower shaded cube having side $BC = v$. To carve the cube with side $x$ from the cube with side $u$, we need to take three slices. Each slice individually has volume $u^2 v$, but they overlap. Each overlap has volume $uv^2$, but the overlaps also overlap in the little shaded cube of volume $v^3$. Thus $x^3 = u^3 - \{(\text{volume of slices}) - [(\text{volume of overlap}) - (\text{volume of overlap of overlap})]\} = u^3 - 3u^2 v + 3uv^2 - v^3$. To this expression for $x^3$ we must add $6x = 6u - 6v$ in order to satisfy the original problem.

**Box 2.4** *Continued*

Now we have $x^3 + 6x = u^3 - 3u^2v + 3uv^2 - v^3 + 6u - 6v$. To make this take the form of a perfect cube, we need to get rid of the "cross terms" $-3u^2v + 3uv^2 + 6u - 6v = -3(uv)u - 6v + 3(uv)v + 6u$, here grouped to highlight $uv$. Now if $uv = 2$, then these terms cancel: $-3(uv)u - 6v + 3(uv)v + 6u = -6u - 6v + 6v + 6u = 0$. So to complete the cube, all we need to do is to stipulate that $u$ and $v$ are determined by the relation $uv = 2$, which we are free to do since the two variables $u$, $v$ are being substituted for the one variable $x$, and so $u$, $v$ must be subject to one additional condition connecting them.

The rest is clear sailing: with this condition applied, our cubic looks like $u^3 - v^3 = 20$, with $v = \frac{2}{u}$. Substituting this back in, we have $u^3 - 8/u^3 = 20$. Multiplying both sides by $u^3$ gives $u^6 - 8 = 20u^3$, or $u^6 = 20u^3 + 8$ if (like Cardano) we want to keep everything positive. This looks worse, because it seems to be of sixth order in $u$, but it is not because it is quadratic in $u^3$. That is, let $y = u^3$; then this equation becomes $y^2 = 20y + 8$. We know how to solve this quadratic equation:

$$y = (20 \pm \sqrt{400 - 4(-8)})/2 = (20 \pm \sqrt{4(108)})/2$$
$$= 10 \pm \sqrt{108} = 10 + \sqrt{108},$$

where the negative choice of sign is neglected because it leads to a negative root, which Cardano would not have accepted. So $u^3 = 10 + \sqrt{108}$ and $v^3 = \sqrt{108} - 10$. Taking the cube root of each side gives $u = \sqrt[3]{\sqrt{108} + 10}$ and $v = \sqrt[3]{\sqrt{108} - 10}$. Finally, recalling that $x = u - v$, we get Cardano's final answer, $x = \sqrt[3]{\sqrt{108} + 10} - \sqrt[3]{\sqrt{108} - 10}$.

sides of the initial square. This allowed Ferrari and Cardano to treat square-square problems in terms of square problems. Here there was no dispute about priority; Ferrari succeeded on his own, though doubtless guided by Cardano's example.

Box 2.5 shows Ferrari's solution in modern notation. We can treat this as a complicated problem of factoring, to use the modern term. Because it followed so quickly on the solution of the cubic, it must have seemed that all such problems were now ripe for solution.

---

**Box 2.5**
Ferrari's solution to the quartic equation

Take the equation $x^4 + a_3 x^3 + a_2 x^2 + a_1 x + a_0 = 0$ and shift the terms so that it becomes $x^4 + a_3 x^3 = -a_2 x^2 - a_1 x - a_0$. Then add $\frac{a_3^2}{4} x^2$ to both sides to make the left-hand side a perfect square:

$$x^4 + a_3 x^3 + \frac{a_3^2}{4} x^2 = \left( x^2 + \frac{a_3 x}{2} \right)^2$$

$$= \left( \frac{a_3^2}{4} - a_2 \right) x^2 - a_1 x - a_0 \qquad (2.5.1)$$

If the right-hand side were a perfect square, we could take the square root of both sides and reduce the equation to a quadratic. In general, it is not a perfect square, but we can make it so by adding to both sides the term $y\left( x^2 + \frac{a_3 x}{2} \right) + \frac{y^2}{4}$:

$$\left( x^2 + \frac{a_3}{2} x + \frac{y}{2} \right)^2 = \left( \frac{a_3^2}{4} - a_2 + y \right) x^2 + \left( -a_1 + \frac{a_3 y}{2} \right) x$$

$$+ \left( -a_0 + \frac{y^2}{4} \right). \qquad (2.5.2)$$

We can make the right-hand side a perfect square by adjusting $y$. This is equivalent to requiring that, in general, $Ax^2 + Bx + C = (ex + f)^2$, where $A, B, C$ are the expressions in parentheses on the right-hand side of equation (2.5.2). If

**Box 2.5** *Continued*

this is to hold, then we must have $A = e^2$, $B = 2ef$, $C = f^2$ or

$$e = \sqrt{A}, f = \sqrt{C} = \frac{B}{2e}.$$

To make this consistent requires that $B^2 - 4AC = (2ef)^2 - 4e^2 f^2 = 0$. So (substituting the values for $A, B, C$ from eq. (2.5.2) into $B^2 = 4AC$), $y$ must satisfy

$$\left(-a_1 + \frac{a_3 y}{2}\right)^2 = 4\left(\frac{a_3^2}{4} - a_2 + y\right)\left(-a_0 + \frac{y^2}{4}\right), \qquad (2.5.3)$$

which is a cubic equation in $y$ (called the "resolvent" of the quartic equation) and hence solvable by the del Ferro-Cardano-Tartaglia method. Once we have determined $y$, then we can use (2.5.3) to determine $e$ and $f$. Then we can return to (2.5.2), now written as

$$\left(x^2 + \frac{a_3}{2}x + \frac{y}{2}\right)^2 = (ex + f)^2, \qquad (2.5.4)$$

which is a perfect square whose square root yields two quadratic equations,

$$x^2 + \frac{a_3}{2}x + \frac{y}{2} = ex + f,$$

$$x^2 + \frac{a_3}{2}x + \frac{y}{2} = -ex - f, \qquad (2.5.5)$$

where $e, f, y$ are now determined. These two equations can readily be solved by the quadratic formula, yielding the four solutions of the quartic equation.

In going past the cubic, however, the obvious references to two and three dimensions disappear and the problem becomes more abstract. How can the human mind understand what it cannot see? As Cardano puts it, "it would be very foolish to go beyond this point. Nature does not permit

---

**Box 2.6**

An illustration of the algebraic notation of Cardano and Viète

Cardano's notation is "syncopated," meaning that it is basically abbreviated words. He does not have a general symbolism for unknowns (which he calls *rebus*, "things") or coefficients.

He expresses his solution to the specific cubic equation in box 2.4 (which he states as "cubus p̄. 6. rebus æqualis 20") as follows:

℞. v. cub. ℞. 108. p̄. 10. m̄. ℞. v. cub. ℞. 108. m̄. 10.

Here p̄. and m̄. stand for plus and minus, and ℞. stands for "radix" (square root), while ℞. v. cub. stands for "cube root." Compare this with its modern form:

$$x = \sqrt[3]{\sqrt{108} + 10} - \sqrt[3]{\sqrt{108} - 10}.$$

In contrast, Viète writes "$A$ cubus + $B$ plano 3 in $A$, aequari $Z$ solido 2" for the modern $x^3 + 3B^2x = 2Z^3$. He uses vowels for unknowns, consonants for coefficients, "cubus" or "solido" for "cubed," and "plano" for "squared"; thus, "$A$ cubus" is $x^3$, "$B$ plano 3" is $3B^2$, and "$Z$ solido 2" is $2Z^3$. Note that he uses the modern + and − signs and also the sign $\ell$ to denote the radical $\sqrt{}$.

---

it." Moreover, the manner of expression of the algebraists left much to be desired. Box 2.6 shows the notation of two Renaissance algebraists, compared with the modern form. The earlier manners of expression operated on entirely different principles; they were primarily a means of abbreviating ordinary language, like shorthand. Modern symbolic mathematics steps decisively away from ordinary language, allowing systematic manipulations that are foreign to natural language. Although we can translate an equation into words, we cannot manipulate those words as we can the equation. However impressive the achievements of the Italian mathematicians, they lacked the insights that would be achieved by symbolic mathematics.

Modern mathematical notation depends on the innovations of a seventeenth-century French lawyer, royal counselor, and codebreaker, François Viète (figure 2.2), who was

**Figure 2.2**
François Viète.

able to merge ciphers, legal abstractions, and mathematics to devise the basic algebraic symbolism we still use. Viète considered himself a rediscoverer of a lost and ancient art, for he could not bring himself to believe that the ancient mathematicians could have achieved what they did with the tools that they claimed to be using. This is not unreasonable; anyone who has read one of Euclid's magnificent geometric proofs cannot help wonder what gave him the idea to draw the crucial lines on which a proof depends. Certainly any student of geometry who has tried to create an original proof understands that there is no "royal road" (as Euclid put it) to finding them. The student may wander down many blind paths and suffer much frustration before hitting on a trick that works.

This is what Pappus had called "synthetic" mathematics. By this he meant that Euclid begins with certain axioms and shows how they can be synthesized to yield the theorem to be proved. But Pappus also alluded to a process of "analytic" mathematics, in which the mathematician would work backward from the desired result, finding on the way what was necessary to arrive at the result in question. There are some tantalizing examples of this working-backward in Apollonius' seminal work on the conic sections, and Viète studied them carefully. He concluded that this analytic art was probably the way the ancients had devised their miraculous proofs: to reach a certain theorem, work backward until you find out what steps are necessary to reach that theorem, then turn around and begin again from that starting point, setting out the steps that you have found to be necessary. Thus, analytic problem solving sets up a kind of scaffolding from which the perfected architecture of the proof can be constructed. At the end, Viète inferred, the ancient masters would remove the scaffolding so that the beholder could admire the beauty of the proof, unobstructed by the tools that built it.

In Viète's view, the ancients obscured traces of their working methods both to excite more admiration and to accentuate the beauty of the finished product. He thought he was only rediscovering their methods, not innovating. However, there is no evidence that the ancients really had the art that Viète reconstructed; and if they did, it was most likely a heuristic working-backward that lacked the symbolism he created. Not knowing the place-system or the Hindu-Arabic zero, the ancients had not taken the decisive steps toward true symbolism. Viète himself was guided by his work in code-breaking, so that the process of solving an algebraic problem through symbolic manipulation became, for him, closely tied to the methods used to solve ciphers, another practice unknown to the ancients. He also may have been influenced by Roman law, which would use a letter to symbolize an unknown defendant (a "John Doe," in modern terms). If one could use such a symbol to represent a person, why not use a symbol to stand for the *cosa*, the "thing" sought for in a problem?

Further, Viète realized that such a symbol might itself be manipulated as if it were a number, rather than an alphabetic letter. Here we have the crucial moment in which algebraic symbolism steps beyond the bounds of ordinary language and semantics, if only through conciseness: compare the brevity of $4x^2$ with "four times the unknown squared." But the advantage of algebraic symbolism goes further: we can add two equations (or otherwise manipulate them) in ways that are impossible for two sentences. What language accomplishes with syntax, algebra can do by juxtaposing a few symbols. Moreover, the ambiguities of language are eliminated in algebra, whose symbols are unequivocal. Where language calls for judgment and logic, algebra requires only symbolic manipulation according to fixed rules. We might even say that the symbols do the thinking, because they embody logical and mathematical power without the user's needing to reinvent it each time, or even understand it beyond following the rules. As our story unfolds, it will become increasingly clear that this power has complex consequences, for it enables amazing discoveries while hiding their meaning behind the very symbolism that discloses them.

Viète called his new discovery a "logic of species," as opposed to a "logic of numbers." He meant by "species" the way the unknown could stand not just for a single number but for a whole class of possible values. That is, "the unknown" is not just a stand-in for a single value but represents the whole spectrum of possibilities: it is a *variable*, as we now say. Here again language falls behind, for Plato had suggested that each word aspires to a natural connection with what it represents, at least for nouns as naming words. Algebra is more like a discourse composed solely of pronouns and verbs and is self-consciously artificial: there is no natural relation between whatever we label the unknown, whether $x$ or $\omega$, and the variable it represents. Unlike language, which is meaning-laden and intentional, algebra removes all trace of ordinary intentionality and meaning. It points toward a very different view of discourse than Plato's, toward a stream of purely conventional symbols, devoid of intrinsic sense.

The abstractness of mathematics is amplified by Viète's even more remarkable discovery: the coefficient. It may not have been so startling to symbolize the unknown by a single symbol, for the Italians were already symbolizing the complex notion of "a number I am looking for but don't know" by the word *cosa* or even by a single character (sometimes even as simple as a dot, ·). But that unknown was always multiplied by a definite quantity, and it was bold of Viète to allow that quantity to become as indefinite as the unknown itself. Here we have the birth of the coefficient, which enables a huge increase in generality and power. Compare the particular case $x^2 - 3x + 2 = 0$ with $a_2 x^2 + a_1 x + a_0 = 0$, which includes the specific equation as a single special case among an infinite expanse of possibilities in which the coefficients $a_2$, $a_1$, $a_0$ roam over all possible values. Indeed, an expression like $a_1 x$ is awkward to render in ordinary language: "Multiply an arbitrary number called $a_1$ by another arbitrary number called $x$." Algebraic symbols avoid the debacle of Babel, bypassing the human confusion of tongues.

Viète not only made these important advances but also grasped their immense power. Though he calls himself merely a restorer of what was lost, he wraps himself in a heroic mantle. He considers Arabic algebra "tedious" and "barbaric," rejecting even their word "algebra" for what he calls instead "the analytic art." Viète rebukes them, as if they were false alchemists, whereas he has the royal touchstone that will actually separate mathematical gold from dross. Viète was cold to the bitter religious controversies of his age. He saves his excitement for mathematics, which, through his newly systematic codebreaking, he uses to protect France. He exults in these analytic powers, realizing that he has purified an ancient art and given it new life.

At the height of those powers, Viète accepted a challenge that pointed to a new frontier in algebra. In 1593, the Belgian mathematician Adriaan van Roomen challenged "all the mathematicians of the whole world" to solve a monster equation of the forty-fifth degree (box 2.7). The Dutch ambassador, paying a call on Henry IV, Viète's royal master, ironically offered the king his condolences that there were no French mathematicians up to the challenge. Stung, Henry called for Viète, who sat down with the problem and was able within a few minutes to find the positive roots of the equation.

**Box 2.7**
Adriaan van Roomen's test problem

Solve

$$x^{45} - 45x^{43} + 945x^{41} - 12,300x^{39} + 111,150x^{37} - 740,459x^{35}$$
$$+ 3,764,565x^{33} - 14,945,040x^{31} + 469,557,800x^{29}$$
$$- 117,679,100x^{27} + 236,030,652x^{25} - 378,658,800x^{23}$$
$$+ 483,841,800x^{21} - 488,494,125x^{19} + 384,942,375x^{17}$$
$$- 232,676,280x^{15} + 105,306,075x^{13} - 34,512,074x^{11}$$
$$+ 7,811,375x^{9} - 1,138,500x^{7} + 95,634x^{5} - 3,795x^{3}$$
$$+ 45x = K,$$

where $K = \sqrt{1\frac{3}{4} - \sqrt{\frac{5}{16}} - \sqrt{1\frac{7}{8} - \sqrt{\frac{45}{64}}}}$.

Viète published this result in 1595, along with his explanation of his method. Humorously, he offered to sacrifice a hundred sheep to celebrate the occasion, ironically echoing Pythagoras' mythical gesture, though with more tranquil animals. Viète emphasized the connection between his "new algebra" and geometry, noting that earlier mathematicians were not able to reconcile them. "But was that not because algebra was up to that time practiced impurely? Friends of learning, embrace a new algebra; farewell, and look to the just and the good." Viète's triumph confirmed his sense that what he had solved was not just one problem but "the proud problem of problems, which is: TO LEAVE NO PROBLEM UNSOLVED."

# 3    Impossibilities and Imaginaries

Van Roomen was so impressed by Viète's astonishing feat that he traveled to France expressly to meet him. Here was a man who claimed to have found the master art that would solve any algebraic problem. Was it really possible that Viète had the tools to solve every possible equation, of whatever degree? In the case of van Roomen's equation of the forty-fifth degree, Viète was lucky, for he recognized that particular equation as closely related to formulas he found in trigonometry. Had van Roomen changed a single coefficient, Viète's solution would no longer have been valid. Clearly, van Roomen had not been envisaging the general problem of solving equations of higher degrees but had come upon that specific equation in the course of his trigonometric studies, and was testing to see if others had discovered it also.

Thus, the solution of van Roomen's equation did not after all illuminate the general question of whether all equations are solvable in radicals, and indeed Viète's solution involved trigonometry, which (as we shall see) goes beyond radicals. The tenor of Viète's conviction was clearly that all equations are solvable by the same general procedures used to solve the quadratic, cubic, and quartic equations. Other

8 Chapter 3

mathematicians were skeptical about algebraic solutions, and their reservations were an important countertheme to the triumphal march of modern algebra. For all his boldness in astronomy, Johannes Kepler was one who remained critical of the claims of algebra, holding that "nothing is proved by symbols, nothing hidden is discovered in natural philosophy through geometrical symbols." This went beyond Kepler's vehement declaration that he hated all kabbalists and their esoteric symbolism. He considered algebra to be "ingenious and surprising," but always thought it fell short of the certainty of geometry, his mathematical touchstone.

The question about algebra comes to a head in Kepler's *Harmony of the World* (1619), in which he tried to place music on a new geometrical foundation, in place of the old Pythagorean numerology of simple whole numbers. The new foundation would be based on the ratios between the sides of the simplest regular polygons. One of his goals was to create a theory that would encompass contemporary musical scales (which used intervals like 5:4 and 6:5), rather than simply the Pythagorean scales (which only used numbers up to 4). To do this, he needed to use all the regular figures up to and including the hexagon but excluding the heptagon, a seven-sided figure that would have introduced harsh dissonances like 7:3 into his scheme. He had a geometrical argument for this, in that the regular heptagon cannot be constructed with straightedge and compass. He was aware, however, that the Swiss mathematician and clock-builder Jost Bürgi had found algebraic equations whose solutions gave the sides of regular polygons. In the case of the heptagon, the equation is cubic and thus solvable by the del Ferro-Cardano-Tartaglia method (box 3.1). This would have caused a crisis for Kepler's musical theory, since accepting this solution would open the door to those terrible dissonances. But the solution involves factors

---

**Box 3.1**

If we call the side of the polygon $x$ and let $y = x^2$, Bürgi's equation for the pentagon is $y^2 - 5y + 5 = 0$, and for the heptagon, $y^3 - 7y^2 + 14y - 7 = 0$. These are both solvable, but the solution of the heptagon's equation using the del Ferro-Cardano-Tartaglia method yields

$$x = \sqrt{\frac{1}{3}\left[\sqrt[3]{\frac{7}{2}}\left(\sqrt[3]{1 - \sqrt{-27}} + \sqrt[3]{1 + \sqrt{-27}}\right)\right]}.$$

---

like $\sqrt[3]{1 - \sqrt{-27}}$, which seemed at that time bewildering if not absurd. Kepler anticipated correctly that the solution of such complex roots would require an infinite process. Since even the "Omniscient Mind" of God could not know such roots in a "simple eternal act ... , neither can it be known by the human mind."

Because geometry always mirrors the "simple eternal act" of the divine intelligence, Kepler preferred it to the questionable processes required to solve cubic equations. The issue was important, for on it hung not just musical questions but the whole structure of the universe, which he took to rest on musical relations. Kepler's search for cosmic harmony led to his Third Law of planetary motion: The square of the period of a planet is proportional to the cube of its mean orbital radius. If the cosmos were organized geometrically, as Kepler believed, algebra might be irrelevant or inadequate. Others paid even less heed. Galileo Galilei never once mentioned algebra, though it would seem likely that he had at least heard of it. When he famously described the "Book of Nature" as written in "the characters of mathematics," he

explicitly meant triangles and geometrical figures, not algebraic symbols. Perhaps here Galileo was influenced by the
academic mathematics of his time, which stood aloof from
the commercial aspects of early algebra.

Nevertheless, Viète's discoveries drew wide attention in
France. His optimism about the possibility of solving every
problem was shared by his great successor, René Descartes,
who was seven when Viète died in 1603 and whose seminal book *La Geometrie* (1637) would consolidate and extend
Viète's results. Despite the closeness of their work, Descartes
was for some reason reluctant to acknowledge Viète's priority, mentioning him only grudgingly and claiming to have
made the same discoveries independently, before reading the
older mathematician's books. Nevertheless, it must be said
that Descartes employed the new algebra with unparalleled
clarity and force. He also put the new discoveries at the service of a new philosophy, reaching toward a larger vision of
understanding the world as matter whose motion is mathe-
matically intelligible.

The thread of algebraic mathematics runs through
Descartes's work. Symbolic mathematics and rigorous rules
exemplify his watchword: method. Indeed, *La Geometrie* and
essays on optics and meteorology form an appendix to his
*Discourse on the Method of Rightly Conducting the Reason.* These
technical works vindicated his insistence on correct method
(*Rules for the Direction of the Mind,* as an earlier work called
it). He applied this touchstone by systematically doubting
the received learning of his time. Descartes also applied his
criteria of method to bring algebra essentially to its modern
form. Viète had stipulated that all the terms in an algebraic
equation should have what we would call the same dimension; that is, if $x$ has the dimensions of length and an equation contains an $x^3$, all the other terms must also have the

dimension of volume (length cubed). Descartes showed that this was unnecessary: we can make all terms have the same dimension by multiplying by a unit length as many times as needed, that is, if 6 and $x$ each have the dimensions of length, we can simply multiply by a unit length to give $6x$ the dimension of volume. Looking at a page of *La Geometrie*, one sees equations that are almost exactly like ours (figure 3.1).

As he shows how the geometry of Euclid and the conic sections of Apollonius can be described by quadratic equations, Descartes illuminates nagging problems that had shadowed these equations. Already the Arabs had realized that some quadratic equations seem to have solutions that are not positive numbers. They did not recognize negative numbers as numbers, although they did seem to have had ways of dealing with debits and credits that approach our concept of signed numbers. Here again the primal notion of number shows its tenacity, for the concept of a "negative number" seems to violate our intuitive sense of counting: how can one count negations? Even Viète did not allow his unknowns or coefficients to be negative. Only in 1629 did Albert Girard recognize negative roots, explaining that "the negative in geometry indicates a retrogression, where the positive is an advance"—essentially the modern idea of a "number line," in which positive and negative represent directions forward and backward along the line. Still, the newcomers were not fully welcome. Other mathematicians of the time refused to admit them, naming them "absurd numbers," and even Descartes called them "false or less than nothing." As late as the eighteenth century, some textbooks denied that the product of two negative numbers is a positive number, a rule that Pierre Simon Laplace said "presents some difficulties" even in 1795.

LIVRE TROISIESME.                   373

ne peut estre diuisée par vn binóme composé de la quau-
tité inconnue ╼ ou ╾ quelque autre quantité, cela tes-
moigne que cete autre quantité n'est la valeur d'aucune
de ses racines. Comme cete derniere

$$x^4 - 4x^3 - 19xx + 106x - 120 \infty 0$$

peut bien estre diuisée, par $x - 2$, & par $x - 3$, & par
$x - 4$, & par $x + 5$; mais non point par $x +$ ou ╾ aucune autre quantité. ce qui monstre qu'elle ne peut auoir
que les quatre racines 2, 3, 4, & 5.

On connoist aussy de cecy combien il peut y auoir de
vrayes racines, & combien de fausses en chasque Equation. A sçauoir il y en peut auoir autant de vrayes, que
les signes ╼ & ╾ s'y trouuent de fois estre changés ; &
autant de fausses qu'il s'y trouue de fois deux signes ╼,
ou deux signes ╾ qui s'entresuiuent. Comme en la derniere, a cause qu'aprés ╼ $x^4$ il y a ╾ $4x^3$, qui est vn chan-
gement du signe ╼ en ╾, & aprés ╾ 19 $xx$ il y a ╼ 106 $x$,
& aprés ╼ 106 $x$ il y a ╾ 120 qui sont encore deux autres
changemens, on connoist qu'il y a trois vrayes racines; &
vne fausse, a cause que les deux signes ╾, de $4x^3$, & 19 $xx$,
s'entresuiuent.

De plus il est aysé de faire en vne mesme Equation,
que toutes les racines qui estoient fausses deuienent
vrayes, & par mesme moyen que toutes celles qui estoiët
vrayes deuienent fausses : a sçauoir en changeant tous
les signes ╼ ou ╾ qui sont en la seconde, en la
quatriesme, en la sixiesme, ou autres places qui se
designent par les nombres pairs, sans changer ceux
de la premiere, de la troisiesme, de la cinquiesme
& semblables qui se designent par les nombres
                    Aaa 3              impairs.

*Marginal notes:*

Cóment on peut examiner si quelque quantité donnée est la valeur d'vne racine.

Combien il peut y auoir de vrayes racines en chasque Equatió.

Cóment on fait que les fausses racines d'vne Equation deuienēt vrayes, & les vrayes fausses.

Figure 3.1
From René Descartes, *La Geometrie* (1637) (p. 373).

Though Descartes refers to "false roots," he does concede that they exist and are roots. He even constructs an ingenious way to tell how many positive and negative roots any equation has, known as "Descartes's rule of signs." It is amazingly simple: "An equation can have as many true [positive] roots as it contains changes of sign, from $+$ to $-$ or from $-$ to $+$; and as many false [negative] roots as the number of times two $+$ signs or two $-$ signs are found in succession." Box 3.2 shows an example, though Descartes does not offer an argument for his rule. Here the power of the new notation comes to the fore, because merely by looking at the signs of the successive terms in the equation one can determine the signs of the roots, even if we don't know their numerical values. Descartes's rule is perfectly general and applies to equations of any degree. He also shows that "it is also easy to transform an equation so that all the roots that were false shall become

---

**Box 3.2**
Descartes's rule of signs

In the page shown in figure 3.1, Descartes explains that the equation $x^4 - 4x^3 - 19x^2 + 106x - 120 = 0$ has three positive ("true") roots and one negative ("false") root because there are three changes of sign between the terms. Consider $x^2 - 5x + 6 = (x - 2)(x - 3) = 0$, which has two changes of sign and two positive roots ($x = 2$ or $x = 3$). If we multiply this equation by $(x - 4)$, we now have three positive roots ($x = 2$, 3, or 4) and three changes of sign in $(x - 2)(x - 3)(x - 4) = x^3 - 9x^2 + 26x - 24 = 0$: each additional positive root can add one more change of sign, at most. Thus, in the page shown in figure 3.1, Descartes explains that the equation $x^4 - 4x^3 - 19x^2 + 106x - 120 = (x - 2)(x - 3)(x - 4)(x + 5) = 0$ has three positive ("true") roots and one negative ("false") root because there are three changes of sign between the terms.

true roots, and all those that were true shall become false," by changing the signs of every other term, in accord with his rule. Surely his awareness that "true" roots could be changed into "false" and vice versa must have helped him accept both as different but comparable kinds of solutions.

Descartes also opens the door for an even more problematic set of solutions that includes square roots of negative numbers, which we still hold at arm's length by using the loaded term "imaginary numbers." As difficult as it is to understand what a negative number might mean, an imaginary number is more perplexing still, and many educated people even today would be hard pressed to account for them. Their alien status is emphasized by calling all non-imaginary numbers "real numbers." The Arabs simply did not allow the square root of a negative number; in such cases, al-Khwārizmī says "there is no equation." Cardano thought such a quantity was "sophistic" and "as subtle as it is useless." Indeed, when one is dealing with quadratic equations, if the root is real, then the formula does not include any imaginary numbers, and all is well. Conversely, if the root is an imaginary number, we could reject it on the grounds that such a solution is not allowable because it is not real. However, in the case of cubic equations, even when all the roots are real, the del Ferro–Cardano–Tartaglia formula *explicitly involves imaginary numbers*. This "irreducible case" (as it came to be known) means that we must come to terms with imaginary numbers if we want to use the formula for the cubic equation. In 1560, Raphael Bombelli had what he called "a wild thought" about how he might resolve this puzzle by showing that the imaginary parts might cancel each other (box 3.3). This trick works in only a few cases, however, leaving open the general question of how to understand imaginary numbers and (even worse) how to take their cube roots or manipulate them when they appear alongside real numbers.

**Box 3.3**
Bombelli's wild thought

Bombelli considered the equation $x^3 = 15x + 4$, whose roots are given, according to the del Ferro-Cardano-Tartaglia solution, by $x = \sqrt[3]{2 + \sqrt{-121}} + \sqrt[3]{2 - \sqrt{-121}}$. Now by simple substitution we can show that $x = 4$ also solves this equation. To reconcile the two solutions, Bombelli proposed the substitution $\sqrt[3]{2 + \sqrt{-121}} = 2 + b\sqrt{-1}$, where $b$ remains to be determined. Cubing both sides of this expression yields $2 + \sqrt{-121} = 2 + 11\sqrt{-1} = (2 + b\sqrt{-1})^3 = 8 + 12b\sqrt{-1} - 6b^2 - b^3\sqrt{-1}$, which is only true if $b = 1$. Similarly, Bombelli showed that $\sqrt[3]{2 - \sqrt{-121}} = 2 - \sqrt{-1}$. The whole solution $x$ is then $x = 2 + \sqrt{-1} + 2 - \sqrt{-1} = 4$. This ingenious trick "seemed to rest on sophistry" to Bombelli; indeed, it cannot be applied to every cubic equation, only special cases. (The solution of the heptagon equation in box 3.1 that perplexed Kepler is one example that cannot be treated in this way.)

Descartes introduced the terms "real" and "imaginary" in their modern sense; in 1831, Carl Friedrich Gauss introduced the term "complex numbers." By giving both parts a name, he granted the imaginary roots a kind of legitimacy, albeit under a name that suggests unreality. This aura persisted even in Gottfried Wilhelm Leibniz's intriguing description of "almost an amphibian between being and nonbeing, which we call the imaginary roots." Leibniz may have been alluding to the mysterious procession of the Holy Spirit in Christian theology, but his description perpetuates a certain spookiness about quantities that arguably have as much reality as "real" numbers. This impression is perhaps best dispelled by thinking of complex numbers as points on a plane, with

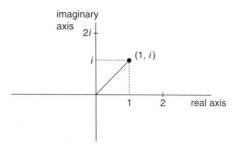

**Figure 3.2**
The two-dimensional representation of complex numbers.

the real part giving the $x$ coordinate and the imaginary part the $y$ coordinate (figure 3.2). In this view, a complex number is fundamentally two-dimensional, whereas the "real" numbers are one-dimensional.

Beyond counting complex roots, Descartes states (but does not prove) that each equation has as many roots as its degree: a quadratic equation has two roots, a cubic three roots, and an $n$th degree equation $n$ roots. This is a crucial insight, originally formulated in 1629 by Girard, which expresses the confidence of Viète and Descartes that all algebraic problems have solutions. It will later lead to the Fundamental Theorem of Algebra, which states that all equations have at least one root, but for the moment that is only a gleam in Descartes's eye.

Descartes also notes that it is possible to shift the roots of any equation by a certain value, even if those roots are unknown. If you wish to increase the value of each root by 3, simply substitute $y - 3$ into the given equation wherever $x$ occurs; the new equation in powers of $y$ will, by definition, have roots $y = x + 3$, shifted as required. To scale the value of each root by a factor $s$, substitute $sy$ for $x$ everywhere;

the new equation in $y$ has roots $y = x/s$. By such simple transformations, any given equation can generate a family of similar equations, with different specific values of the roots in each case. However, these transformations do not affect the *relative* relation of the roots, which are all affected equally. We will return to this "relativity" of roots (which Descartes did not discuss) in chapter 9.

Descartes is so sure that equations of degree beyond four can be solved that he doesn't bother to give specific procedures, but instead offers a "general rule": try to simplify the equation to one of lower degree by factoring it. Not only does he have the work of Cardano and Ferrari to rely on, but he (like Viète before him) has been able to use his analytic approach to geometry to solve problems that had baffled the ancients: the trisection of an angle, the duplication of the cube. To be sure, in those cases he has to use tools not allowed in ancient geometry, that is, not just straightedge and compass but also conic sections. Descartes's proudest achievement is to have solved another problem that troubled the ancients, the four-line locus problem posed by Pappus: to find the curve that lies at a given distance from each of four given lines. Not only does Descartes solve this problem with almost arrogant ease, but he goes on to solve a five-line locus problem, which the ancients did not even attempt. These triumphs occupy him far more than the problem of solving equations beyond the quartic. He does give one example, solving an equation of the sixth degree by ingenious manipulations of conic sections, but his equation is not completely general; he restricts its coefficients to allow his geometric method to succeed. With an ironic flourish, he concludes that "it is only necessary to follow the same general method to construct all problems, more and more complex, ad infinitum; for in the case of a mathematical progression, whenever the first two or three terms are given, it is easy to find the rest." He asks pardon for what he has "intentionally omitted, so as to leave to others the pleasure of discovery."

But having left the problem of equations beyond the fourth degree as an exalted exercise for the reader, he evidently does not realize how difficult that exercise will prove to be. The next centuries contain many attempts to realize the vision of Viète and Descartes. Though there had been extraordinary progress in solving cubic and quartic equations during the sixteenth century, it is perhaps not so surprising that the issue of the quintic equation lay unresolved for two hundred years more. After Descartes, a whole huge field of investigation lay open, in which the question of solving higher-degree equations was only one problem among many. Yet this strand is richly interwoven in the mathematical history of the two centuries after Descartes, which will produce the greatest mathematicians yet to have lived: Newton, Euler, and Gauss.

# 4    Spirals and Seashores

Sir Isaac Newton's relation to algebra was peculiarly ambivalent. As a young man, he studied Descartes's works carefully. In maturity, he so loathed Descartes that he sometimes seems to have been unwilling to utter or write Descartes's name, as if it would defile him. The reasons for this are not totally clear. Newton despised Descartes's philosophy as cloaked atheism, but there were also mathematical reasons for his displeasure. Newton, like Viète, styled himself a devotee and restorer of antiquity, whereas Descartes, he felt, had betrayed the profundity of Greek geometry for the questionable advantages of analytic geometry.

To demonstrate his claim, in his *Principia* Newton offers a purely geometric resolution of the four-line locus problem Descartes so prided himself on, noting caustically that this gives "not an [analytical] computation but a geometrical synthesis, such as the ancients required, of the classical problem of four lines, which was begun by Euclid and carried on by Apollonius." In the rest of the *Principia*, Newton also tends to avoid analytical algebra, preferring to state his propositions in the manner of Euclid. However, this appearance is deceptive, for in fact Newton does use algebraic expressions as well as a new tool of analytical mathematics that he himself has created: the calculus. Though he phrases it in terms of geometry, Newton's calculus goes beyond anything known to the ancients, even Archimedes, whom Newton considered his precursor.

Thus, though he tends to prefer geometry, Newton is steeped in algebra. As a young professor, he lectures on the topic (during 1672–1683), though his lectures on *Universal Arithmetic* are not published until 1707. Newton also makes a number of discoveries that will bear on the problem of the quintic. He was able to make simple connections between the coefficients of an equation and its roots, known as "Newton's identities." These generalize "Girard's identities": the coefficient of the next-lowest degree term (of $x^4$, for a quintic equation) is equal to the negative sum of all the roots (see box 4.1). Likewise, all the other coefficients are equal successively to the sum of all products of the roots taken two at a time, then three at a time, until the final, constant term is equal to the negative of the product of all the roots. Later on, it will become extremely important to note that all of the roots appear in a *symmetric* manner in each of these products and sums. That is, every root appears in exactly the same way as every other root, so that if one were to exchange two roots, the product of all of them would be unchanged, as would the products of them taken two at a time, and so on. The importance of these rules is that they let us see direct relations between coefficients and roots, without knowing the value of the roots. Newton is also able to obtain upper and lower bounds for the roots, that is, to show how large (or small) they could possibly be. Using these tools, we can look at any equation and determine the range within which its roots lie, as well as whether they are negative or positive in value.

Newton's deepest insight, however, remained hidden for many years, buried in a passage in his *Principia* that was

Box 4.1
Girard's and Newton's identities

Consider a cubic equation with the roots $x_1$, $x_2$, $x_3$. The equation can be written $(x - x_1)(x - x_2)(x - x_3) = 0$. Multiplying it out, we get $x^3 - (x_1 + x_2 + x_3)x^2 + (x_1x_2 + x_1x_3 + x_2x_3)x - x_1x_2x_3 = 0$. Note that the coefficient of the $x^2$ term is the negative sum of all three roots, $-(x_1 + x_2 + x_3)$, while the coefficient of the $x$ term is the symmetric product of all the roots, taken two at a time: $(x_1x_2 + x_1x_3 + x_2x_3)$. Finally, the constant term of the equation is the negative product of all three roots: $-x_1x_2x_3$. We can apply the same reasoning to an equation of any degree, so that the coefficient of the next-to-highest power of the unknown must be the negative sum of all the roots, the next coefficient must be the symmetric sum of all the roots taken two at a time, and so on. We will refer to these simple relations as "Girard's identities," which Newton greatly generalized in expressions he derived for the sum of the square of all the roots, or the sum of their $n$th powers, "Newton's identities."

much noticed until the twentieth century. In lemma 28 of Book I, in an investigation of the curved paths that bodies can take (as in the orbits of planets), he shows that "No oval figure exists whose area, cut off by straight lines at will, can in general be found by means of equations finite in the number of their terms and dimensions." A line $ax + by + c = 0$ crossing an oval cuts off an area that cannot be expressed by a polynomial function of $a$, $b$, and $c$. Though he does not define the word explicitly, by "oval" Newton seems to mean any closed curve that does not cross itself (it is "simple," in the language of modern mathematicians) and is infinitely smooth (it always has finite curvature, never is "flat"). The simplest such curve is a circle, and it had long been suspected that the area of a circle is irrational with respect to its radius. But what Newton surmises goes far beyond the irrationality of $\pi$, the name later given (by Euler) to the value of a circle's perimeter divided by its diameter, and beyond the irrationality even of $\pi^2$. Newton's argument indicates that the area of a circle is not given by *any* algebraic function, however high its degree, and thus that area (and hence $\pi$ also) cannot be expressed in terms of *any* finite number of square roots, cube roots, fifth roots, and so on.

To use the term that Euler later introduced, the area of a circle is *transcendental*, meaning it cannot be expressed as the root of any equation of finite degree whose coefficients are rational numbers. At one stroke, Newton indicates that such magnitudes exist (because circles exist, and have areas), and also that there are infinitely many of them, since his proof is not restricted to circles but holds for any "oval" curve. His proof is a miracle of simplicity and power, for which he does not even bother to draw a picture or write down a line of algebra. It follows from a single brilliant contrivance. Inside the oval, pick any point whatever; let us call it the pole, $P$. Now let a straight line come out from that pole and rotate around it at uniform angular speed. Picture a clock hand that makes a complete revolution in one hour. Now imagine a point of light moving along that hand, starting from the pole and moving outward along the hand with speed given by the square of the distance from the pole to the point $A$ where the hand intersects the oval (figure 4.1).

Newton has set up a way of measuring the area of the circle, for each hour the hand sweeps through that area, and the moving point keeps track of the area because it is traveling with speed proportional to the area swept out. Here Newton is implicitly using his new calculus of motion, for he knows

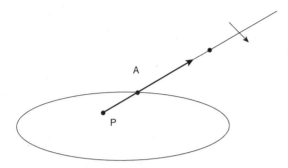

**Figure 4.1**
Newton's diagram for his lemma 28, which argues that no oval curve has an area expressible by a finite algebraic equation. From any point $P$ inside the oval, draw a straight line that rotates about $P$ at uniform angular speed.

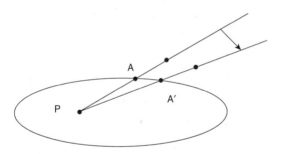

**Figure 4.2**
Detail of Newton's lemma 28; the speed of the moving point is proportional to the area swept out between $A$ and $A'$.

that, in an infinitesimally short time, the point travels a distance from the pole equal to the area the hand has swept out in that time (figure 4.2). However, we don't need to know anything about calculus in what follows. All that matters is that since, second by second, the moving point is registering the area that the hand sweeps out, we can measure the whole area of the oval merely by waiting until an hour has elapsed and measuring the distance the moving point has traveled radially outward from the pole. For the two-dimensional problem of measuring an area, Newton has substituted a one-dimensional problem that gives the same answer: find the length traveled outward by the moving point during an hour.

The hand moves around uniformly, but the moving point speeds up and slows down in the course of each hour, in proportion to the square of the distance from the pole to the oval at any given moment. Each hour it returns to its initial speed of an hour before. If you were to watch the lighted point (or if you were to open the shutter of your camera and make a long exposure), you would see it move in a spiral, starting at the pole and making "an infinite number of gyrations," as Newton puts it (figure 4.3). Now Newton applies a *reductio ad absurdum:* Suppose that it is possible to describe this spiral (and hence also the area of the oval) by some polynomial equation with a finite number of terms, $f(x, y) = 0$. Then consider a straight line running across the spiral that we will call the $x$-axis, defined by $y = 0$. What can be said about the intersections of this line and the spiral? Each of them is a root of the equation $f(x, 0) = 0$. For instance, Descartes showed that all the conic sections can be described by equations of the second degree, and those curves can be cut by a straight line no more than two times. Now Newton relies on the fact that an equation of finite degree can have only a *finite* number of roots, no matter how large. But the spiral in its "infinite number of gyrations" crosses the line an infinite number of times. Thus, there should be an infinite number of intersection points, corresponding to an infinite number of roots of the equation. This contradicts our hypothesis that the equation has finite degree, and hence Newton's conclusion follows: there is no such equation that gives the area of the oval.

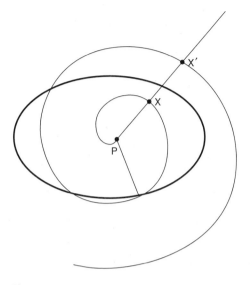

**Figure 4.3**
In Newton's lemma 28, the track of the moving point forms a spiral, composed of the motion of the point along the line and the uniform rotation of the line itself about $P$. Newton determines the area of the curve by comparing the distance from the pole $P$ to the point $X$ after one full revolution, which sweeps out the full area of the oval. The moving point travels an equal distance $XX' = PX$ during the next sweep, and so forth.

This brilliant argument indicates that all simple closed curves, such as the circle or the ellipse, have areas that cannot be described by finite algebraic equations. The argument seemed so simple that it made Newton's contemporaries suspicious. Daniel Bernoulli and Leibniz tried to state counterexamples, but these each involved a curve that intersects itself (for example, the lemniscate, a figure-8 on its side, ∞) or that is not closed (for example, a parabola). Later mathematicians demanded greater rigor than Newton's beautifully simple arguments (how can we *prove* that the spiral must have

infinitely many intersections with the line?), but the basic
thrust of his insight was sustained. There are infinitely many
magnitudes that are more irrational than any radical, in the
sense that no finite root is commensurable with them. In this
sense, they are transcendental. Implicitly, Newton considers
that his proof argues the priority of geometry over algebra
by showing that a simple geometric figure includes quantities that defeat any finite amount of algebra. This makes his
own preference for geometry (and his geometrically phrased
calculus) more persuasive, for he thereby places his master
theory beyond the limitations of algebra. In Newton's view,
the ancients, with him as their modern champion, have de-
feated the upstart Descartes by subsuming algebra under the
larger umbrella of geometry.

With this in mind, it is understandable that Newton may
not have considered the solution of the quintic to be germane
to his larger project, though in his younger days he did spend
much energy in classifying cubic equations and contributing
to the advance of algebra (for instance, "Newton's method"
for the approximation of roots). If the real battle concerns
geometric magnitudes not expressible in finite equations,
why worry about the details of quintics? Ironically, succeeding generations of mathematicians would take up Newton's
work in the simpler algebraic notation devised by his arch
rival, Leibniz. His insights about transcendental magnitudes
would be rediscovered centuries later. We will return to them
later in light of the unfolding story of the quintic.

Despite Newton, the power and beauty of the algebraic
notation ensured that interest in basic questions of algebra
remained high. New hope for the solution of equations
beyond the quartic was offered by the work of a Saxon noble-
man, Count Ehrenfried Walter von Tschirnhaus. Tschirnhaus
had varied interests. He served in the Dutch army and spent

time in Paris and England. His interest in optics led him to
set up glassworks in Italy. He was called "the discoverer of
porcelain" because he helped set up the pottery works at
Dresden (though the art had been long known in China). In
algebra, he sought to generalize Cardano's approach to the
cubic by setting up transformations that would simplify a
higher-degree equation. In 1683, Tschirnhaus showed that,
in a polynomial of degree $n$ (greater than 2), the terms proportional to the two next-highest powers of $x$ (namely, $x^{n-1}$
and $x^{n-2}$) can always be eliminated with one of his transformations (box 4.2). Thus, in a quintic, we can always change
variables to get rid of the $x^4$ and $x^3$ terms. A century later, the
Swedish mathematician E. S. Bring showed that we can also
get rid of the $x^2$ term, leaving the general quintic in the much
simplified form $x^5 + px + q = 0$. The quintic now looked so
simple that it became increasingly perplexing why it would
not yield. Indeed, George B. Jerrard was convinced that he
had found a solution, defending it as late as 1858.

The quintic's intractability was considered a challenge that
simply awaited the right assault, and it was widely believed
to be solvable until almost 1800. However, already in the
1680s, when Tschirnhaus's friend Leibniz turned his attention to the solution of the quintic, he noticed what later
emerged as evidence of a fatal difficulty. Leibniz wondered
if an extension of Tschirnhaus's technique could simplify the
quintic into the form $x^5$ equals a constant, which would be
readily solvable. Unfortunately, the equations needed to do
this are of *higher* degree than five. Thus, this approach would
not really simplify the problem of the quintic because it leads
to equations that are even more complicated, not less. Leibniz
thus judged that Tschirnhaus's method, at least, was doomed
to fail, though he did not draw the inference that no other
method could succeed.

**Box 4.2**
Tschirnhaus's transformation (1683)

For any equation of the $n$th degree,

$$a_n x^n + a_{n-1} x^{n-1} + \cdots + a_2 x^2 + a_1 x + a_0 = 0,$$

consider a substitution that transforms the roots $x_1, x_2, \ldots, x_n$ into new roots $z_1, z_2, \ldots, z_n$, such that

$$z_1 = b_4 x_1^4 + b_3 x_1^3 + b_2 x_1^2 + b_1 x_1 + b_0,$$
$$z_2 = b_4 x_2^4 + b_3 x_2^3 + b_2 x_2^2 + b_1 x_2 + b_0,$$
$$\vdots$$
$$z_n = b_4 x_n^4 + b_3 x_n^3 + b_2 x_n^2 + b_1 x_n + b_0,$$

where $b_4, b_3, \ldots, b_0$ are functions in radicals of the original coefficients $a_0, a_1, a_2, \ldots, a_n$.

Tschirnhaus showed that we can eliminate the two next-higher powers of the unknown, $x^{n-1}$ and $x^{n-2}$ by choosing $b_4, b_3, \ldots, b_0$ through solving these quartic equations. In 1843, George B. Jerrard showed that the $x^{n-3}$ term could also be removed by a similar transformation. Thus, these transformations allow any quintic equation to be rewritten in the much simpler form $x^5 + px + q = 0$. However, in general this procedure is extremely complex; even if we use a computer, it takes megabytes of memory to store the result.

In this period much attention was also paid to the general question of whether all equations do have at least one solution, which is the Fundamental Theorem of Algebra that Girard and Descartes had stated without real proof. In 1748, Jean Le Rond d'Alembert offered a proof, but it was insufficient, as were the efforts of Leonhard Euler in the years following. Despite the basic simplicity of the theorem, it proved difficult to establish because of the vast generality

of the possibilities. That is, solving an equation can be visu-
alized (following Descartes) as finding the point (or points)
at which a curve $y = x^n + a_{n-1}x^{n-1} + \cdots + a_0$ crosses a given
line (say the $x$ axis, $y = 0$; see figure 4.4). We already know
that this happens for the quadratic, cubic, and quartic equations, for those intersections are just the real roots that were
found by Cardano, Ferrari, and the rest. But how can we
be sure that there isn't even one weird equation that does
not cross the line? Euler was able to give many ways to factor equations of higher degree into products of lower-degree
equations, but, despite his immense ability and productivity

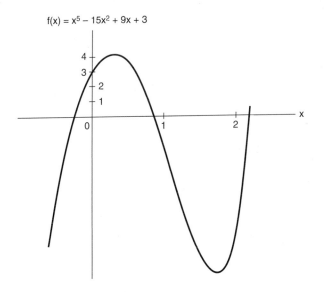

**Figure 4.4**
The graph of a quintic equation, $y = f(x) = x^5 - 15x^2 + 9x + 3$, whose three
crossings of the $x$ axis give the three real roots of the equation $y = 0$. Note
that two of the roots are positive, as given by Descartes's rule of signs (see
box 3.2). It turns out that these roots cannot be expressed in radicals.

(which continued even after he lost his sight), he could not give a completely general result.

That was left for "the prince of mathematicians," Carl Friedrich Gauss. Even before he was eighteen, Gauss had made an important discovery about a special class of equations that correspond to inscribing regular polygons inside a circle. Euclid's geometry only allows the use of straightedge and compass. Descartes showed that, in algebraic terms, this means that solutions of the corresponding equations can involve only square roots. For this reason, it seemed impossible to construct a regular polygon having a prime number of sides greater than five without recourse to higher roots than square roots. The young Gauss showed that, on the contrary, a seventeen-sided polygon could be constructed, meaning that he had found a solution of its special sixteenth-degree equation, expressible solely in terms of square roots. This discovery confirmed his decision to devote himself to mathematics rather than philology.

Gauss was to give four proofs of the Fundamental Theorem over the course of his long career. The first was his doctoral dissertation of 1799. The last (and perhaps the most elegant) allows $x$ to be a complex number, with both a real and an imaginary part, and allows the coefficients also to be complex. The use of complex numbers enables Gauss to ground general statements in a kind of picture. Figure 4.5 shows the heart of his proof. It depicts a quintic equation, but the same considerations will apply to any equation. Gauss represents the behavior of the equation in terms of the polar coordinates $r$ and $\theta$. The figure shows a complex plane, as defined in figure 3.2. In that plane, we have a circle whose radius $R$ is chosen to be so large that the equation's behavior is dominated by the behavior of its leading term, $x^n$ (or $r^n$, in polar coordinates), when $r$ is greater than $R$. Gauss shows that one can

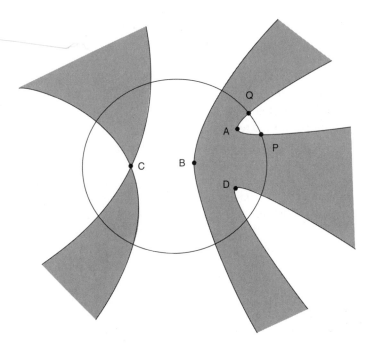

**Figure 4.5**
Gauss's diagram to establish the Fundamental Theorem of Algebra. The shaded areas are the "seas," in which the real part of the polynomial is less than zero; the unshaded areas are "land," in which the real part is greater than zero. Walking along the "shore" from $P$ to $Q$, at some point we must find a root (for instance, at $A$). Likewise, we find roots at $D$ and $C$ (here, a double root).

always find such a circle, no matter what the equation. He also shows that, going around that circle, the real part of the equation takes alternately positive and negative values; as it does so, the imaginary part must also change sign from negative to positive and thus pass through zero.

To complete the proof, Gauss draws lines showing the regions where the real part of the polynomial is positive (unshaded in the figure), separated from the regions where it is negative by lines indicating where the real part is zero. This picture can be understood if we think of the shaded (negative) region as "sea" and the unshaded (positive) region as "land"; the lines show the seashore, where the real part of the polynomial is zero. Now if there is a solution, both the imaginary and the real parts of the polynomial must vanish. So let's take a walk along the seashore from $P$ to $Q$, keeping the "sea" always on our left. From Gauss's earlier result, we know that the imaginary part of the equation must go from being negative to being positive as we walk along the shore. This means that at some point in the walk the imaginary part will be zero, and that is the root, because the real part is zero all along the shore. This is point $A$ in the picture; there are also roots at $B$ and $D$ and a double root at $C$ (that is, two roots that happen to be equal; note the intersecting shorelines there).

Thus, this particular quintic has five roots (two of them equal), and one can make a similar argument for any other quintic, no matter what "geography" the seas and land have, for there will always be shorelines curving in some manner, along which the real and imaginary parts will both vanish at some point. Here Gauss invokes arguments that later came to be called *topological*, meaning that they make use of the overall features of a figure that do not depend on its detailed shape. By his choice of examples, it is clear that Gauss thought he was tracking the solutions of the quintic. He had just shown that they must exist. There they are, right at the water's edge. The problem was simply how to express them.

# Premonitions and Permutations

Gauss's full proof of the Fundamental Theorem was preceded by many inconclusive attempts. Mathematicians anticipated that all equations would have solutions, following the vision of Viète and Descartes, and most were not daunted by the centuries-long roadblock at the quintic. Then, in the late eighteenth century, Joseph-Louis Lagrange brought to light evidence that would prove crucial. In doing this, he revealed an approach that would open the way to a new kind of mathematics. Those who came afterwards regarded his 1771 paper, "Reflections on the Algebraic Theory of Equations," as a foundational document.

Lagrange began by asking why the solutions of the cubic and quartic had succeeded in the first place. He reviewed the process of factoring (completing the square or the cube) and concluded that this sort of procedure could not work for the quintic. Of course, this did not mean that the quintic could not be solved, but only that it was insoluble through this well-tried method.

Assume, then, that we know the three roots of the cubic equation, $a_3x^3 + a_2x^2 + a_1x + a_0 = 0$; we can call them $x_1, x_2, x_3$. As he analyzed these roots, Lagrange noticed that,

> **Box 5.1**
> Roots of unity
>
> We are looking for roots that satisfy $\alpha^3 - 1 = 0$, other than $\alpha = 1$. We can start by factoring out a term $(\alpha - 1)$, yielding $\alpha^3 - 1 = (\alpha - 1)(\alpha^2 + \alpha + 1) = 0$, as can be verified by multiplying it out. For $\alpha \neq 1$, we must then have $\alpha^2 + \alpha + 1 = 0$. This is a simple quadratic equation whose solution is $\alpha = \frac{-1 \pm \sqrt{-3}}{2}$, which are the other two cube roots of unity besides 1. The equation for the $n$th root of unity, $\alpha^n - 1 = 0$, can be factored in a similar way, yielding $\alpha^n - 1 = (\alpha - 1)(\alpha^{n-1} + \alpha^{n-2} + \cdots + \alpha^2 + \alpha + 1) = 0$. If $\alpha \neq 1$, then $\alpha$ must satisfy $\alpha^{n-1} + \alpha^{n-2} + \cdots + \alpha^2 + \alpha + 1 = 0$.

in the solution of cubic equations, a special role is played by the three cube roots of unity, which are the solutions of the simple equation $\alpha^3 = 1$ (see box 5.1). These are solutions of the simplest possible cubic equation and thus give the basic elements of the solution of more complicated equations. This is not surprising. In the case of the quadratic equation, the solution of $\alpha^2 = 1$ gives $+1$ and $-1$, the two fundamental elements from which all other (positive and negative) numbers are built. Gauss's work, mentioned in the preceding chapter, also showed the general importance of the equation $\alpha^n = 1$ in constructing an $n$-sided regular polygon. Clearly, the solutions of this equation would also be important in finding the solution of an $n$th degree algebraic equation.

Returning to the general cubic equation, Lagrange then took the three roots $x_1, x_2, x_3$ and formed a very simple expression: $R = (1x_1 + \alpha x_2 + \alpha^2 x_3)^3$, the cube of the sum of the products of each of the three roots times each of the three cube roots of unity. At this point, Lagrange noted a simple and striking property. Our naming the roots $x_1$, $x_2$, $x_3$ is quite arbitrary; we could just as well switch the names around and call them $x_2$, $x_1$, $x_3$ or $x_3$, $x_2$, $x_1$ without changing anything. What this means is that we can *permute* the roots in various ways. In fact, there are just $6 = 3 \times 2 \times 1$ different ways to permute the 3 roots. In general, given $n$ quantities, there are $n! = n \times (n-1) \times (n-2) \times \cdots 2 \times 1$ permutations, where $n!$ is read "$n$ factorial" (see box 5.2).

By examining the permutations of the roots, Lagrange found a crucial clue to understanding the solution of the equation that will be all-important in what follows. (The point was made independently in 1770 by Alexandre-Théophile Vandermonde.) Lagrange asked what happens to $R$ when we permute the roots. As box 5.3 shows, although there are six different permutations of the three roots, they lead to only two possible values for $R$. Lagrange thought this was "very remarkable" and used it to show how a cubic equation can be reduced to the solution of a quadratic (box 5.4).

Lagrange extended this idea to higher-degree equations: There is always a quantity analogous to $R$ that takes a certain

---

**Box 5.2**
Permutations

Consider $n$ objects, for instance the 52 cards in a deck. There are $n$ ways to pick the first one of them (52 for a deck). For each of these, there are $(n-1)$ ways to pick the second (51), hence $n(n-1)$ ways to pick the first two ($52 \times 51$). Likewise, there are $(n-2)$ ways to pick the third, ..., all the way to 1 way to pick the final item. Thus there are $n! = n \times (n-1) \times (n-2) \times \cdots \times 2 \times 1$ permutations.

**Box 5.3**
Lagrange's permutation of the roots of a cubic equation

Consider $R_1 = (1x_1 + \alpha x_2 + \alpha^2 x_3)^3$ and make the substitution $x_1 x_2 x_3 \xrightarrow{(123)} x_2 x_3 x_1$, which can be written symbolically as (123). Under this substitution, $R_1 \xrightarrow{(123)} R_2 = (1x_1 + \alpha x_3 + \alpha^2 x_1)^3$. Under the substitution $x_1 x_2 x_3 \xrightarrow{(132)} x_3 x_1 x_2$, or (132), $R_1 \xrightarrow{(132)} R_3 = (1x_3 + \alpha x_1 + \alpha^2 x_2)^3$. Of the six different permutations of the three roots, these three are even permutations (counting the possibility of leaving them unpermuted, the "identity"), meaning that an even number of transpositions have been made (two for $R_2$ and $R_3$, zero for the identity). There are three odd permutations left: (32), which means interchange $x_2 \leftrightarrow x_3$, leaving $x_1$ alone, yielding $R_4 = (1x_1 + \alpha x_3 + \alpha^2 x_2)^3$; (12), yielding $R_5 = (1x_2 + \alpha x_1 + \alpha^2 x_3)^3$; and (13), yielding $R_6 = (1x_3 + \alpha x_2 + \alpha^2 x_1)^3$. However, these expressions turn out to be related: $R_2 = R_1$, $R_3 = R_1$, $R_5 = R_4$, and $R_6 = R_4$. As a result, there are only two distinct values for $R$ under these permutations, which we call $t_1 = R_1 = R_2 = R_3$ and $t_2 = R_4 = R_5 = R_6$.

number of values when the roots are permuted, and this number is an important indicator of solvability. In the case of the quartic, Lagrange formed a resolvent $R$ from the four roots (box 5.5). These roots can be permuted in $4! = 24$ different ways, but under these 24 permutations $R$ takes on only three values, the three roots of a cubic, and from there Lagrange could re-derive Ferrari's solution.

Thus, in all the soluble equations so far, when the roots are shuffled around, the resolvent $R$ takes on *fewer* values than the degree of the equation. This reflects the fact that the equation can be reduced to solving one of lower degree. The cubic's resolvent has two values and the quartic's resolvent

**Box 5.4**
Lagrange's resolvent for a cubic equation

Following box 5.3, consider again the cubic equation $x^3 + a_2 x^2 + a_1 x + a_0 = 0$. Girard's identity (box 4.1) $-a_2 = x_1 + x_2 + x_3$ and the definitions $\sqrt[3]{t_1} = \sqrt[3]{R_1} = 1x_1 + \alpha x_2 + \alpha^2 x_3$, $\sqrt[3]{t_2} = \sqrt[3]{R_4} = 1x_1 + \alpha^2 x_2 + \alpha x_3$ provide three *linear* equations that determine $x_1, x_2, x_3$ in terms of $\sqrt[3]{t_1}, \sqrt[3]{t_2}$, and the constant $a_2$. To find $t_1$ and $t_2$, Newton's identities allow us to calculate $t_1 + t_2 = -2a_2^3 + 9a_1 a_2 - 27a_0$ and $t_1 t_2 = \left(a_2^3 - 3a_1\right)^3$. Then we can express $t_1$ and $t_2$ as the two roots of the quadratic equation $(t - t_1)(t - t_2) = 0 = t^2 - t(t_1 + t_2) + t_1 t_2$ or

$$t^2 + t\left(2a_2^3 - 9a_1 a_2 + 27a_0\right) + \left(a_2^3 - 3a_1\right)^3 = 0,$$

the "resolvent equation" mentioned in the text. The final result is identical to the del Ferro-Cardano-Tartaglia formula.

has three, each one smaller than the degree of the equation being permuted. When Lagrange turned to the quintic, he found that the resolvent takes on six values, a *larger* number than the degree of the equation. This showed that the method that worked to solve equations up to quintic was breaking down. This was also realized independently by the Italian Gianfrancesco Malfatti (1770) and may have been noted by Leibniz much earlier, as mentioned in the previous chapter.

Lagrange drew from this work the moral that a new way of attacking the quintic was needed, and he probably hoped that his work would enable him to find it, if only by steering him away from the false paths of the past. But he was not yet close to being convinced that the quintic is not solvable. What he did argue is that "it is very doubtful that the methods we have just discussed can give a complete solution of equations

**Box 5.5**
Lagrange's resolvent for the quartic equation

For the quartic equation $x^4 + a_3 x^3 + a_2 x^2 + a_1 x + a_0 = 0$, consider the four roots $x_1, x_2, x_3, x_4$ and the function $R = (x_1 + x_2 - x_3 - x_4)^2$. As we go through the 24 permutations of the 4 roots, $R$ takes on only 3 distinct values:

$$R_1 = (x_1 + x_2 - x_3 - x_4)^2,$$
$$R_2 = (x_1 + x_3 - x_2 - x_4)^2,$$
$$R_3 = (x_1 + x_4 - x_2 - x_3)^2.$$

As in the case of the cubic, the four roots can be expressed as simple combinations of these different values of $\sqrt{R}$, namely $\sqrt{R_1}$, $\sqrt{R_2}$, $\sqrt{R_3}$, and the condition that the sum of all four roots equals $-a_3$. In turn, various sums and products of the resolvent can be expressed in terms of the coefficients of the original quartic equation, using Newton's identities. From these expressions, Lagrange showed that $R$ must satisfy a cubic equation. Having solved this cubic equation, whose three roots are $R_1$, $R_2$, $R_3$, we can recover $x_1$, $x_2$, $x_3$, $x_4$, which are exactly as given by Ferrari's formula.

of degree five and, all the more so, of higher degrees." In the process of working out the resolvent for the quintic, he noted that the computations are "so long and complicated that they can discourage the most intrepid calculators," by which he means human calculators, in that pencil and paper era. For him, all this reflected an "uncertainty [that] will discourage in advance all those who might be tempted to use [the older methods] to solve one of the most celebrated and important problems of algebra."

Thus Lagrange still remained hopeful that new methods might succeed where the old ones failed. His contemporary, the mathematical historian Jean Étienne Montucla, expressed this optimistic spirit in a military metaphor, as if the quintic were a town under siege: "the ramparts are raised all around but, enclosed in its last redoubt, the problem defends itself desperately. Who will be the fortunate genius who will lead the assault upon it or force it to capitulate?"

Here Gauss came to a different conclusion. In his doctoral dissertation of 1799, he first recorded his belief that the quintic equation has no general solution in radicals. Gauss announced this possibility along with his first proof of the Fundamental Theorem of Algebra. In 1801, Gauss published his opinion in his masterwork, *Disquisitiones Arithmeticae*, a compendium of many remarkable results. Among them, Gauss included his famous work on constructing regular polygons, which involved the question of the solvability of their equations. It was natural that, in this context, his thoughts turned to the more general problem of solvability. "Everyone knows that the most eminent geometers have been ineffectual in the search for the general solution of equations higher than the fourth degree. ... And there is little doubt that this problem does not so much defy modern methods of analysis as that it proposes the impossible." Gauss, knowing the extent of his own powers, would have been the last to give up had he not divined far deeper problems than those his ingenuity and persistence might solve.

But while Gauss guessed this new level of difficulty, he did not explain *why* the quintic is unsolvable. He never seemed to be attracted to permutations and combinations, so it is not clear that his opinion was founded on the kind of permutational arguments that Lagrange had made. All the more striking, then, that even before Gauss announced his opinion, a little-known mathematician in Italy formulated and sought

to prove the assertion that the quintic is unsolvable in radicals. The son of a physician, Paolo Ruffini was born in a
small Italian town, moving in childhood to Reggio near the
city of Modena (figure 5.1). As a child, he seemed destined
for the priesthood, but he finally chose to study philosophy,
medicine, and surgery, graduating in 1788. He served as a
professor of mathematics in Modena until the upheavals in
the wake of the French Revolution hit Italy. Though not a
revolutionary himself, Ruffini refused to swear a loyalty oath
to the French and was barred from his university post until
1799. During this enforced absence from academic life, his
mind was preoccupied with mathematics, though the bulk
of his time was devoted to medicine. At the end of the day, he
would jot down his thoughts and ideas on letters he received
and in the margins of papers he read. His thoughts turned to
the theory of equations as he prepared his first work, which
included his first proof of the impossibility of solving the
quintic.

Ruffini's devotion to medicine was deep, despite his mathematical preoccupations. In 1817 to 1818, there was an epi-
demic of typhoid fever, during which Ruffini tended to the
sick with great dedication and courage. He himself fell ill
with the disease. Characteristically, he used his experience
to write a paper on contagious typhoid. He also maintained
his philosophical and religious interests, writing about the
definition of life and against Laplace's mechanistic theory
of morality. His contemporaries noted his self-effacing modesty and spirituality. Indeed, in later life Ruffini was offered
a mathematical post at the University of Padua, but declined
because he was reluctant to leave his many patients.

Perhaps because of this modesty, Ruffini has been somewhat neglected, despite his deep merits and the singular
interest of his dual life as physician and mathematician.

**Figure 5.1**
Paolo Ruffini.

Though he did not seem to covet the usual rewards of professional glory, however, Ruffini was tenacious in his quest to prove the unsolvability of the quintic and to bring his work before the mathematical world. He published no fewer than six versions of his proof (the first in 1799, the last in 1813), several of them responding to questions and criticisms raised both by his friend Pietro Abbati and by the older mathematician Malfatti, who could not accept the notion of unsolvability.

He began his first proof by announcing that "the algebraic solution of general equations of degree greater than 4 is always impossible. Behold a very important theorem which I believe I am able to assert (if I do not err); to present the proof of it is the main reason for publishing this volume. The immortal Lagrange, with his sublime reflections, has provided the basis of my proof." Indeed, Ruffini was among the first to grasp the immense importance of Lagrange's idea of "permutations" (as Ruffini called them) and was the first to offer a proof of the unsolvability of the quintic.

Unfortunately, Ruffini expressed these proofs in long and obscure forms, which rendered them hard to understand. He sent his proof to Lagrange, hoping for recognition from the man whose work had given him his inspiration. Lagrange never replied, a bitter disappointment to Ruffini and his friends, for they considered Lagrange (born in Turin, though of a French family) as a fellow countryman who they thought should recognize this Italian achievement. Later, Lagrange was on a committee to examine Ruffini's memoir, but he was cool toward it, finding "little in it worthy of attention." As it later emerged, Lagrange (then elderly) may have had trouble wading through the paper and eventually told a third party that it was "good" but did not give sufficient proof of some of its assertions. Indeed, he may not have spent much time studying Ruffini's proof.

Other mathematicians were more favorable. Some readers at the Royal Society were "quite satisfied" with Ruffini's proof. Most important, the great French mathematician Augustin Cauchy recognized the importance of Ruffini's work and generalized some of its results during 1813 to 1815 (as will be discussed later and in appendix C). One can imagine Ruffini's gratification when, six months before his own death, he received a letter from Cauchy, who wrote that "your memoir on the general resolution of equations is a work that has always seemed to me worthy of the attention of mathematicians and that, in my judgment, proves completely the unsolvability of the general equation of degree greater than 4." Cauchy also told Ruffini that he had cited this work in his lectures.

Despite Cauchy's strong endorsement, Ruffini's work still did not achieve general acceptance among mathematicians. The reasons are unclear but seem to center on the novelty of arguments from permutations and even more on the radical notion that some equations might not, in general, be solvable in radicals. This might have been behind Lagrange's wariness, for he described Ruffini's proof as a "difficult matter," something that he himself thought improbable, if not incredible, and for which he thought Ruffini's reasoning insufficient. To be sure, later mathematicians confirmed Lagrange's contention that there was an important missing step in Ruffini's arguments. Thus Ruffini's proof fell short, at least in the sense that it left others unconvinced. He deserves credit for his vision and his courage, but it remained for others to resolve the issue decisively.

# Abel's Proof

Ideas, especially mathematical ideas, have a life that goes beyond their human creators. Without knowing Ruffini's work, Niels Henrik Abel (figure 6.1) discovered the same argument only a few years later and also gave the first essentially complete demonstration. Abel's story is one of the most moving in the history of mathematics. In his twenty-six years of life, he discovered whole new territories in mathematics, though struggling constantly with poverty and misunderstanding. His native Norway was then in its youth as an independent nation, and Abel's checkered fortunes to some extent mirrored Norway's struggle.

When Abel was 12, Norway separated from Denmark, long the dominant partner in their dual kingdom, and set up her own parliament, the Storting. But Norway could not completely separate from her more powerful neighbors. In 1814, Abel's father was among the delegates sent to offer the crown of Norway to the reigning Swedish king, Karl XIII. Abel's father was a pastor, like his father before him, and had a notable career as a member of the Storting. He was a man of the Enlightenment who read Voltaire and was active in the movement for literacy and vaccination in rural Norway. He wrote several popular catechisms and books of prayer

**Figure 6.1**
Niels Henrik Abel.

doubtless read by his son. He was interested in natural science and woke his children to see lunar eclipses. However, his political career ended ignominiously when he pressed unfounded accusations against certain powerful persons. He died an alcoholic, leaving nine children and a widow who also turned to alcohol for solace. After his funeral, she received visiting clergy while in bed with her peasant paramour.

Abel, then eighteen, found himself without support and obliged to act as the responsible adult for his younger siblings. Somehow, though, he continued his education. He had been fortunate to find a mentor in a young mathematics teacher, Berndt Michael Holmboe, who inspired him and became his lifelong friend. Abel soon showed an amazing ability to solve difficult problems. While still a high school student, he read Lagrange's work and Cauchy's 1815 paper on permutations. (Though Cauchy's paper was based on Ruffini's work, Abel did not read Ruffini, perhaps because his works were hard to find and in Italian.) Holmboe recognized and extolled his student's "excellent mathematical genius." In 1821, Abel entered the Royal Frederick's University, which had opened in 1819 in the new capital, Christiana (now called Oslo), then a city with only 11,000 inhabitants. He continued to make rapid strides, beginning to write original papers and going far beyond the skill of his teachers. With admirable understanding, those teachers proposed that he be granted a special fellowship to visit Paris and Berlin, though in fact he had little to learn even from the greatest mathematicians of the time.

While always modest, Abel during this time set himself to solve the most difficult and famous problems. In 1823, he tried his hand at Fermat's celebrated last theorem, to show the impossibility of finding integers $a$, $b$, and $c$ that would satisfy the equation $a^n = b^n + c^n$, where $n$ is an integer greater than 2. Not surprisingly, he found himself "at the end of my tether," as he put it, for this problem resisted solution until 1993, when it finally succumbed to very elaborate abstract techniques. Even so, he found what now are called "Abel's formulas," which showed that if any solutions existed, they would have to be extremely large numbers. Abel also turned his mind to the problem of the quintic equation. At first, he thought he had managed to solve it and was very excited. This was in 1821, twenty-two years after Ruffini had published the first version of his proof. At this point, Abel still did not know Ruffini's work, which is not surprising considering the isolation of Norway and its lack of mathematical libraries. Indeed, during the winter months, the Oslo fjord remained frozen and mail was often delayed.

Though Abel had probably taken note of Gauss's opinion that the quintic was unsolvable, he nonetheless refused to give up the search for a solution. After all, Gauss offered no proof for his assertion, and Abel, like Lagrange and most other mathematicians, could well consider that the search was not over. However, when his teachers asked him to give some numerical examples of his solution to the quintic, Abel soon realized that it was not as general as he had thought. This was a decisive moment, for he could have stubbornly resumed the quest for a fully general solution to repair the gaps in his work, on the assumption that a solution *had* to exist. Instead, he made an about-face and turned his efforts to proving the unsolvability of the quintic. He never explained his reasons, and one wonders what moved him. In 1824, two years after Ruffini's death, Abel published his first proof of the unsolvability of the quintic, which in many ways is close to Ruffini's proof, although it fills in an important gap that Ruffini had not noticed. In 1826, Abel read anonymous articles summarizing the work of Ruffini, which Abel acknowledged in his final (posthumous) paper: "The first, and if I am not mistaken the only one, who before me had tried to show the impossibility of the algebraic solution of general equations is the geometer *Ruffini*. But his paper is so complicated that it is very difficult to decide the correctness of his reasoning. It seems to me that his reasoning is not always satisfactory."

Because Ruffini's work overlaps largely with what Abel went on to do, I will not discuss it separately. I do not mean to belittle the recognition Ruffini is due. The theorem of unsolvability may better be called the Abel-Ruffini Theorem. Yet Abel's remark, as well as Lagrange's hesitation, indicate aspects of decisive proof in which Ruffini fell short. A proof that lacks a decisive step is not yet a proof. Accordingly, I will summarize Abel's version, indicating along the way the common elements in their arguments and the crucial gap in Ruffini's reasoning that Abel filled.

If you wish to read Abel's own words, appendix A contains a translation of his 1824 paper. Abel had this earliest version of the proof printed at his own expense as a pamphlet, hoping to use it as a "calling card" that would gain the attention of the great mathematicians, Gauss above all. However, to save paper and money, Abel compressed his argument to telegraphic terseness, which he amplified in his later accounts of the proof. Accordingly, I have added a running commentary to help the reader follow Abel's argument. Appendixes B and C fill out details he merely mentions. Here I will present the four crucial stages of the proof as Abel presented them in 1826, without spelling out all the technicalities to be found in the appendixes. In each case, I will present a summary statement, followed by my explanation.

Abel uses the time-honored method of *reductio ad absur-dum*: he begins by assuming that the quintic is solvable and shows that this leads to a contradiction. His first step is to specify the form that a solution must have. Given a general equation of the $m$th degree (taking $m$ as a prime number, so that it cannot be factored further),

$$a_m y^m + a_{m-1} y^{m-1} + a_{m-2} y^{m-2} + \cdots + a_2 y^2 + a_1 y + a_0 = 0, \quad (6.1)$$

Abel proves a very general statement about any solution, which he calls an "algebraic function":

(I)    *All algebraic functions y can be expressed in the form*

$$y = p + R^{\frac{1}{m}} + p_2 R^{\frac{2}{m}} + \cdots + p_{m-1} R^{\frac{m-1}{m}}, \qquad (6.2)$$

*where $p$, $p_2$, ... are finite sums of radicals and polynomials and $R^{\frac{1}{m}}$ is in general an irrational function of the coefficients of the original equation.*

That is, if $y$ is the solution of an algebraic equation of degree $m$, $y$ can be expressed as a series of terms that contain nested roots involving the coefficients and irrational expressions like $R^{\frac{1}{m}}$. (Remember that $R^{\frac{1}{m}}$ is just another way of writing the $m$th root of $R$, $\sqrt[m]{R}$.) This is very close to the general form that Euler had already conjectured some years before. Abel gives a detailed proof (see appendix B, including some subtleties neglected here), but the idea can readily be illustrated with a few examples. As box 6.1 shows, the solution of the quadratic equation can be expressed as $y = p + R^{\frac{1}{2}}$, which is just the general form above with $m = 2$ and $p$ a simple polynomial. Likewise, the solution of the cubic can be expressed as $y = p + R^{\frac{1}{3}} + p_2 R^{\frac{2}{3}}$, which follows the general form, with $m = 3$. In both of these cases, $m = 2$ and $m = 3$, $m$ is a prime number. For the quartic, $m = 4$ is not prime and the general solution

**Box 6.1**
Abel's form for the quadratic equation

In the quadratic equation, $y^2 - a_1 y + a_0 = 0$, substitute $y = p + R^{\frac{1}{2}}$ to yield $(p^2 + R - a_1 p + a_0) + (2p - a_1)R^{\frac{1}{2}} = 0$. To satisfy this in general, each parenthesis must be separately zero (as Abel discusses in [A8] in appendix A), so $p = \frac{a_1}{2}$ and then $R = -a_0 + \frac{a_1^2}{4}$. Thus, $y = \frac{a_1}{2} \pm \frac{1}{2}\sqrt{a_1^2 - 4a_0}$, the familiar form of the quadratic solution. For the case of the cubic solution, see p. 174 in appendix B.

involves only combinations of the quadratic and cubic forms, as was discussed earlier.

In the case of the quintic, Abel's general form (6.2) becomes

$$y = p + R^{\frac{1}{5}} + p_2 R^{\frac{2}{5}} + p_3 R^{\frac{3}{5}} + p_4 R^{\frac{4}{5}}. \tag{6.3}$$

Because this follows from his general result about the form of the solution, either the solution of the quintic has this form, or there is no such solution. So Abel assumes hypothetically that the quintic does have a solution of exactly this form. He now goes on to show that this form leads to a contradiction. This requires three further steps.

The next step is crucial; Ruffini had assumed it without giving a proof, but Abel remedies this.

(II) *All algebraic functions y can be expressed in terms of rational functions of the roots of an equation.*

The idea here is simple but telling. In equation (6.3), the general form of $y$ is expressed in terms of polynomials and also various irrational functions (here, $R^{\frac{1}{5}}$, $R^{\frac{2}{5}}$, $R^{\frac{3}{5}}$, $R^{\frac{4}{5}}$) of the *coefficients*. But Abel brilliantly proves that we can express

---

**Box 6.2**
The relation between roots and coefficients

Let $y = p + R^{\frac{1}{2}}$, as shown in box 6.1, which also shows that $R = -a_0 + \frac{a_1^2}{4}$. Now consider the two roots, $y_1 = \frac{a_1}{2} + \frac{1}{2}\sqrt{a_1^2 - 4a_0}$, $y_2 = \frac{a_1}{2} - \frac{1}{2}\sqrt{a_1^2 - 4a_0}$. Then $(y_1 - y_2) = \sqrt{a_1^2 - 4a_0}$ and thus $(y_1 - y_2)^2 = a_1^2 - 4a_0 = 4\left(-a_0 + \frac{a_1^2}{4}\right) = 4R$. The name *discriminant* is given to $a_1^2 - 4a_0$, which is zero when the roots are equal, $y_1 = y_2$, positive when the roots are real, and negative when they are imaginary.

---

$y$ in terms of the *roots* instead of the original *coefficients* of the equation. What is important here is that all the various *irrational* functions of the coefficients that appear in $y$ (e.g., $R^{\frac{1}{2}}$) are *rational* functions of the roots of the equation. To choose a simple example, in the quadratic equation above, whose general solution is $y = p + R^{\frac{1}{2}}$, box 6.2 shows that $4R = (y_1 - y_2)^2$, where $y_1$, $y_2$ are the two roots of the quadratic. So $2R^{\frac{1}{2}}$, the square root of $4R$, is equal to the difference of the two roots, $(y_1 - y_2)$. In agreement with Abel's proof, this is indeed a simple rational function of the roots.

In the case of the quintic, Abel's step II implies that $R^{\frac{1}{5}}$ is a rational function of the roots, as are its powers, $R^{\frac{2}{5}}$, $R^{\frac{3}{5}}$, $R^{\frac{4}{5}}$, as well as $p, p_2, \ldots$. Because it is made up of products of all these rational functions, $y$ is thus a rational function of the roots. Here there is an interesting tension: The roots are irrational functions of the *coefficients*, but those coefficients are always rational sums or products of the *roots*. This harks back to Girard's identities (see box 4.1), which showed that the coefficients were sums and products of the roots.

Let us return to the tension between the irrationality of the roots as functions of the coefficients, on the one hand, and the rationality of the coefficients as products of the roots, on the other. Roughly speaking, as the degree of the equation grows higher, it is harder and harder to reconcile this tension. Finally, with the quintic equation, it is no longer possible, and we cannot, in general, find a solution in radicals. However, this idea must be amplified much further.

The next step limits the hypothetical solution in a way that will prove to be decisive.

(III) *If a rational function of five quantities takes fewer than five values when the five quantities are permuted, it can take only two different values (equal in magnitude and opposite in sign), or one value, but never three or four values.*

Abel drew this theorem from the work of Cauchy, of which it is a special case. Cauchy's proof is presented in appendix C. It relies on looking at the different ways we can permute the roots of the equation. Abel also relies on the Fundamental Theorem of Algebra: A quintic equation must have at least one root and no more than five different roots. So there can be no more than five values for $y$, which, according to step II, is a rational function of the roots of the equation.

As before, let us consider any of the expressions in the solution that is a rational function of the roots, such as $R^{\frac{1}{5}}$. Cauchy's result requires that $R^{\frac{1}{5}}$ can take only one, two, or five values as the roots are permuted, but never three or four values. This gives Abel leverage to show that the solution cannot work. First, note that $R^{\frac{1}{5}}$ cannot, in general, take only one value, because then it could lead to a single solution, not five roots, which we have assumed to be unequal. Then, Abel investigates what happens if $R^{\frac{1}{5}}$ takes on five values (appendix A, [A15]–[A16]). This leads to a contradiction; he shows that $R^{\frac{1}{5}}$, having five possible values when the roots are permuted, would have to be equal to an expression having 120 values, which is impossible.

That excludes the possibility that there are five values for $y$. So Cauchy's result must require that there be only two (appendix A, [A27]). But this too leads to a contradiction, because when we switch the five roots around, Abel shows that we get an inconsistent result again. He derives an equation whose left-hand side has 120 possible values, while the right-hand side has only 10. Clearly, such an equation cannot be solved in general, and so the hypothetical solution leads to absurdities. Therefore Abel concludes that

(IV) *It is impossible to solve the general equation of the fifth degree in radicals.*

The strategy of Abel's argument is straightforward: he takes the only possible form a solution could have and shows that it leads to contradictory results when we permute the roots of the equation. This contradiction rests on a special property of the number five, shown by the number of values the hypothetical solution can take when subjected to permutations of its five roots. The argument also applies for degrees higher than five. For instance, we can multiply an unsolvable quintic by a factor of $y = y - 0$, and it would then be a sixth-degree equation that has one root $y = 0$ and five unsolvable roots, and similarly for any higher degree.

Though Abel's argument shows the impossibility of solving the quintic in general, it still seems opaque. The question remains: *why* this impossibility? To examine this further means we must seek the heart of Abel's proof.

# Abel and Galois

Abel published his proof shortly before he began his travels abroad, hoping that it would open doors for him. He knew that this trip was a godsend, his only chance to make a mark on the larger world. He was so poor that he barely scraped by at home; only the stipend given him by his university made such a dream voyage possible. Because he was shy and prone to loneliness, he decided to travel with friends, fellow Norwegians also pursuing studies abroad. For this reason he did not go directly to Paris, the mathematical center of the world at that time, but headed rather to Berlin. He hoped to meet Gauss, but the great man received very few visitors, turned inward on his own concerns. He received Abel's proof but did not cut its pages, setting it aside as if it were from a crank: "Here is another of those monstrosities!"

In Berlin, Abel did meet August Crelle, an amateur mathematician of some distinction who had founded a journal of mathematics that would become so well known that it would be called simply *Crelle's Journal* by its readers. Crelle was hospitable to Abel and soon recognized his new friend's extraordinary gifts. He published seven of Abel's papers in the first volume of his journal and also began working to find him a position in Berlin. For Abel, this attention buoyed his hopes that he might secure a living as a mathematician that would allow him to marry his fiancée, Christine Kemp, who, like Abel, had no means and remained in Norway working as a governess.

Despite the requirement that he spend the bulk of his time in Paris, Abel could not resist accompanying his friends on their journeys to Italy and Switzerland. His letters pleaded with his professors in Norway not to disapprove of his desire to see these fabled places, even though the trip strained his precarious finances to the limit. Though he was then in good health and young, he had a premonition that this would be his only chance to see the world.

At last, Abel took leave of his friends and steeled himself to face Paris alone. There, after all, he could meet Adrien-Marie Legendre and Cauchy, eminent members of the Académie des Sciences, well qualified to appreciate Abel's achievements and promote his career. His reception was chilly. His formal visits to the great men were received politely but without real warmth or recognition. They paid no attention to his proof. They were more absorbed in their own concerns and saw him only as one more bright young man among so many that had come seeking their favor over the years. Abel wrote to a friend that the French were "monstrous egotists ... uncommonly reserved with respect to foreigners. .... Everyone works by himself here, without bothering others. Everyone wants to teach and no one wants to learn." Though Abel was disappointed, he kept working diligently and continued to send Crelle more and more papers, as he spread his wings ever further.

He did make an impression on some Frenchmen, at least. Jacques Frédéric Saigey, the mathematical editor of Baron de Férrusac's *Bulletin*, an important journal of the day, asked

Abel to write brief accounts of articles in other journals, including his unsolvability proof for the quintic. It was at this point (1826) that Abel became aware of Ruffini's work and noted its unsatisfactory quality. Abel also met an important self-taught scientist, François-Vincent Raspail, an ardent partisan of the French Republic who was several times exiled or imprisoned. In his scientific work, he had made steps toward the development of the cell theory of biology and the notion of microbe-borne disease. Raspail was quite struck by Abel.

Despite his loneliness and dwindling funds, Abel found diversion in Paris. He played billiards and indulged his passion for the theater, which he described in animated letters. At the end of 1826, his funds ran out and he had to return home months early. He hoped to find a post in Norway, but the only vacancy in the whole country had been filled during his absence, ironically by his teacher and mentor, Holmboe. Abel was not bitter, and their friendship continued, but he must have been deeply perplexed. He could not think of marrying in such penury; he had to eke out his living as a substitute teacher in a military academy. Nevertheless, he had not stopped thinking about his proof of unsolvability, seeking its larger implications.

In March 1828, he came to a crucial realization. In a new paper he sent to Crelle, Abel wrote that "Although the algebraic solution of equations is not possible in general, there are nevertheless particular equations of all degrees that admit such solutions. Such are, for example, the equations of the form $x^n - 1 = 0$," which Gauss had solved so brilliantly when younger than Abel. "The solution of these equations is founded on certain relations among the roots. I have sought to generalize this method by supposing that two roots of a given equation are so connected that one can express rationally the one by the other, and I have come to the result that such an equation can always be solved with the help of a certain number of equations of *lower degree*. There are even cases where one can solve *algebraically* the given equation itself."

Abel gives this interrelation of the roots a simple form. He calls the first root $x_1$; then he assumes that the next root $x_2 = f(x_1)$, where $f(x_1)$ is a rational function of $x_1$. He then assumes that the next root is the same function of the previous root, $x_3 = f(x_2) = f(f(x_1))$, which he calls $f^2(x_1)$, meaning the function $f$ applied two successive times. He then keeps applying $f$ to each successive root and gets the series of roots $x_1, f(x_1), f^2(x_1), f^3(x_1), \ldots, f^{n-1}(x_1)$, where $n$ is the degree of the equation. For instance, if $n = 5$, $x_6 = f^5(x_1) = x_1$, since the equation can have only five different roots, which one can arrange in a circle,

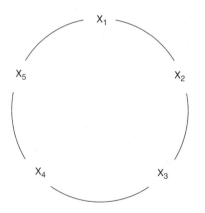

showing the cyclical symmetry of the roots of such *solvable* "abelian equations," as they are now known.

Abel's insight is that, in these solvable equations, any two roots are related by a rational function. If this is true, then all the roots are rational functions of each other and of the coefficients of the given equation. We need one further condition, however, and it is crucial. It is implicit in the case we have considered, but becomes explicit when Abel considers an equation whose degree can be factored into several primes, so that the roots can be grouped into different cycles if the equation is solvable. If $x$ is a root of such an equation and two other roots are given by $f(x)$ and $g(x)$, which are two (possibly different) rational functions of $x$, then Abel concludes that the equation is always solvable if

$$g(f(x)) = f(g(x)).$$

That is, if the *order* in which these two functions are applied to $x$ does not matter, then the equation is solvable.

This is the insight that I consider most helpful. At first, it may seem merely formal or devoid of significance. Abelian equations are solvable if their roots are related by functions such that it does not matter in what order we apply the functions. Contrariwise, the equation may not be solvable if the order does matter. But a great surprise is hidden in this seemingly flat statement. Until this point, none of the basic operations of arithmetic and algebra has been "noncommutative." Thus, $a + b = b + a$ and $a \times b = b \times a$. These so-called commutative laws express an important quality of numbers, and they hold sway in the operations used in every equation, of whatever degree.

Abel's insight connects solvability with commutativity, at least for abelian equations. Thus, he opened the whole issue of commutativity for consideration. Indeed, I believe that this was the first time that the possibility that operations might not commute emerged in mathematics, especially in the context of simple equations where we would least expect such a strange thing. Though Abel did not live to see it, his successors gradually extended and elucidated the subtle connection between noncommutativity and unsolvability. In the rest of this book, I hope to follow this story and explore its implications.

Notice, first of all, how Abel's insight into solvable equations squares with this. He was able to organize the roots of abelian equations into the cyclic pattern shown above. It is clear that in such cases the relations between the roots are commutative, for it does not matter whether we go around the circle backward or forward. Abel emphasizes this commutative symmetry as the crux of this pattern. Yet this insight is only a beginning, for it does not explain how to determine whether the pattern of a given equation is or is not cyclic in this way. Abel must have seen that he needed to take it much further. His 1828 paper leaves us wondering how he might have carried his question about commutativity even further.

Just at this point, however, Abel's brief strand of life ran out. He was ill at the beginning of 1828 and did not know how widely his papers were being read, for Norway still remained isolated. Although he was not given to undue enthusiasm, Gauss now spoke of the "depth, delicacy, and elegance" of Abel's work, and it may be that Abel heard of this through Crelle. He probably did not yet know of the admiration of Legendre or Carl Gustav Jacob Jacobi, who was vying with Abel to extend his results still further. At age twenty-three, Jacobi was already an associate professor at the University of Königsberg. Indeed, both Jacobi and Abel were proposed for membership in the Institut de France in 1828, though neither was elected. Crelle was working hard to find a place for Abel, who described himself as being "as poor as a churchmouse" in a note begging an old friend for a loan, which he signed "Yours destroyed."

Crelle was not the only one trying to help Abel. Four distinguished French mathematicians (including Legendre and Siméon-Denis Poisson) wrote to the king of Sweden imploring some help for one of his own subjects, "a young mathematician, Monsieur Abel, whose works show that he has mental powers of the highest rank and who nonetheless grows ill there in Christiania in a position of too little value for one of his so rare and early-developed talent." The king never responded, and Abel probably never knew of this attempt to intervene on his behalf.

For his part, Abel continued to work, not just on the solvability of equations but also on elliptic functions and what now are called Abelian integrals, making fundamental contributions. He began to receive some letters from Legendre that comforted him by their frank admiration and interest in his work, even though Legendre seemed to have lost an important paper that Abel had sent to the Institut years before (which Abel did not mention in his replies). Generously, Abel wrote back that Legendre's recognition gave him "one of the happiest moments" of his life, "for I would have accomplished nothing without having been led by your light." With childlike pride, Abel quoted such praise to his friends. He also wrote: "I am, however, almost completely alone. I assure you that in the most profound sense I am not in association with a single human being. Nevertheless, this lack of friends is not foremost in my mind because I have so horribly much to do for [Crelle's] *Journal*."

At Christmas, 1828, Abel could not resist the opportunity to spend the holiday with his fiancée in the country, even though the winter was exceptionally cold. He was increasingly ill, but his friends could not dissuade him from going. He set off, with only socks to warm his hands. After the Christmas balls and festivities, he took to his bed with pneumonia. He went on to complete one more paper on tran- scendental functions, in order to record part of the treatise lost in Paris. Crelle wrote him that he was almost sure of having secured him a position in Berlin, which would allow him financial security and the possibility of marriage at last. But the matter was not completely resolved. Abel kept denying that he was dying of tuberculosis. During his last days, he was brave. During the nights, his anger and despair fright- ened those around him. He cursed his poverty and the in- justice of the neglect he had suffered, but still made light of his condition. He died on April 6, 1829, in the presence of his fiancée Christine. Two days later, not knowing of his friend's death, Crelle sent final word that Abel's appoint- ment was approved. On his deathbed, Abel had sent word to his friend Keilhau urging him to marry Christine after his death. Though he had never met her, Keilhau proposed in a letter and they were soon married.

Thus, Abel's new insight of 1828 appeared in print only after his death. We can only speculate where he might have taken these ideas. Yet even before Abel's death, and without knowing his final work, a young Frenchman had already taken the steps that would generalize Abel's insight into a full account of the solvability of equations. The story of Évariste Galois (figure 7.1) stands beside that of Abel among the most dramatic lives in mathematics. Galois's story is, however, far better known, elaborated and romanticized in novels, films, and biographies. Abel died at twenty-six, but Galois died at only twenty years old, in a mysterious duel linked to love and politics.

Galois was born in 1811 in Bourg-la-Reine, just south of Paris. His family had adopted the ideals of the Revolution of 1789, which ended more than eight hundred years of rule by the Bourbon-Capetian dynasty. After Napoleon's first

**Figure 7.1**
Évariste Galois.

downfall and exile to Elba, however, France turned back to
its royalist past, and Louis XVIII took the throne vacated by
his executed brother, Louis XVI. When Napoleon attempted
a comeback from Elba, Galois's father became the mayor of
Bourg-la-Reine, a position he held for fifteen years, even after
Napoleon's defeat at Waterloo in 1815. At age eleven, Évariste
entered the famous Parisian Collège de Louis-le-Grand as a
boarding student. Despite its great name, the school was as
grim as a prison. The students rose at 5:30 A.M. to study in
silence for two hours before a breakfast of bread and water. Often punished, they were put in solitary cells for days
at a time. By now Louis XVIII had returned to power, and
the school's administration was rigidly royalist. Nonetheless, Galois made brilliant progress and won prizes in Greek
and Latin.

Like Abel, Galois was helped by his teachers, first
Hippolyte Jean Vernier and then Louis-Paul-Émile Richard.
Though shy, Richard went out of his way to seek oppor-
tunities for his amazing student. He approached Cauchy,
who presented one of Galois's papers to the Académie des
Sciences in 1829, to be reviewed by himself and other
distinguished members. It was a crushing disappointment
when that judgment never came; indeed, Galois never even
got his manuscript back, for unclear reasons. Galois blamed
Cauchy for losing the manuscript, but some documents suggest instead that Cauchy urged Galois to revise and expand
his work. Galois might well have been somewhat paranoid,
and stories told after his death do reveal a darker side. The
mathematician Sophie Germain describes Galois attending
the sessions of the Académie and insulting the speakers.
The romantic myth of misunderstood genius ignores such
provocative and probably outrageous behavior. Yet the
whole story is sadly reminiscent of Abel's disappointment in

Paris in 1826 to 1827, when his submission to the Académie also disappeared. (Abel described Cauchy as "a Catholic bigot.") Though they were in Paris at the same time, it does not seem that Abel and Galois ever met, though we might imagine them passing each other, unknowing, in the street. Certainly they were pondering the same questions about the solvability of equations in 1828, though Galois was not aware of Abel's work as he formed his own theory.

Those years were terrible for Galois. Louis XVIII's successor, Charles X, was even more desirous of a return to the *ancien régime*, and the clerical party was also gaining ascendancy. In their little town, the Jesuit parish priest forged Galois's father's name to scurrilous epigrams directed to his own relatives. In the ensuing scandal, Galois's father felt forced to leave the town and was so humiliated that he committed suicide in 1829. The grieving Évariste saw the parish priest insulted and hit with stones at his father's funeral. That summer, he took the entrance examinations for the prestigious École Polytechnique. During them, Galois was questioned on logarithmic series. When his examiners asked him to prove his statements, he refused, saying that the answers were completely obvious. His obstinacy ended his dream of being a *polytechnicien*. Through the intervention of Richard, Galois was able to gain admission to the École Préparatoire (now called the École Normale Superieure), where he did not bother to hide his scorn for his teachers and studied only mathematics.

In 1830, rioting began in Paris against the government; Galois wanted to join in the fight but was kept locked up in the school. In July, Charles X, driven from Paris, was unable to regain power. The old General Lafayette, a moderate like his friend George Washington, allowed Charles's dynastic rival, the duc d'Orléans, to become Louis-Philippe

I, the "peoples' king." Galois and many others considered this "July monarchy" a grotesque betrayal of their struggle to restore the Republic. Along with Raspail, Galois joined the radical Société des Amis du Peuple and became more active politically, for which he was finally punished by being expelled from school. His situation then became even more strained. He tried to eke out a living by giving private lessons in mathematics. He sent yet another paper outlining his theory of equations to the Académie, which this time rejected it, infuriating him even further. During a banquet of the Société, Galois raised his jackknife and sarcastically toasted "To Louis-Philippe!" In the commotion, he was not heard as he continued: "if he betrays his oaths."

Though acquitted of the charge of treason, Galois soon was put in prison on other charges, along with Raspail, who later described how Galois, who still looked like a child, kept working on his mathematics even there. Taunted by the other inmates, Galois tried to show he could hold his liquor. On one such occasion, Raspail remembered that Galois had to be restrained from a drunken attempt at suicide. On another, Galois escaped a bullet fired by a guard into his crowded cell.

In 1832, the scandal about the Institut's treatment of Abel came out, and Galois in prison vented his anger about "the very men who already have Abel's death on their conscience," whom he blamed for his own neglect. That year, the republicans were unable to organize further demonstrations, for so many of their leaders were in prison. When an outbreak of cholera struck Paris, the youngest prisoners were transferred on parole to a clinic. There, Galois fell in love with Stéphanie Poterin-Dumotel, the daughter of one of the doctors. Two fragmentary letters indicate that, though initially encouraging, she finally declined to accept his love. To a friend, he wrote: "Happiness and hope are at an end, now surely con-

sumed for the rest of my life." The legend of Galois comes
from the succeeding events. He died only a few weeks after this letter in a duel whose motives and details remain
obscure. Some writers ascribe romantic causes, others political; the full truth will probably never be clear. The surviving
documents contain tantalizing evidence, but do not permit a
decisive solution.

After being released from prison, his love rejected, Galois
wrote: "Pity, never! Hatred, that's all." His final letters contain his most explicit testimony, though it leaves much un-
explained: "I die the victim of an infamous coquette and two
of her dupes." Consider also this account of the duel given
by a newspaper in Lyon: "The young Évariste Galois . . . was
fighting with one of his old friends, a young man like himself,
like himself a member of the Société des Amis du Peuple, and
who was known to have figured equally in a political trial. It
is said that love was the cause of the combat. The pistol was
the chosen weapon of the adversaries, but because of their
old friendship they could not bear to look at one another and
left the decision to blind fate. At point-blank range, each of
them was given a pistol and fired. Only one of the pistols was
loaded."

If this newspaper account is accurate, two comrades in the
same secret political society, both involved with the same
woman, used Russian roulette to decide who would sur-
vive. In his final letters, Galois begged his comrades "not
to reproach me for dying otherwise than for my country,"
indicating that he died for purely personal reasons. Yet the
mystery remains. Galois wrote that "only under compulsion
and force have I yielded to a provocation which I have tried
to avert by every means." He reproached himself with having told "the hateful truth to those who could not listen to
it with dispassion. But to the end I told the truth. I go to the

grave with a conscience free from patriots' blood. I would like to have given my life for the public good. Forgive those who kill me, for they are of good faith." What was the provocation to which he yielded, the hateful truth he spoke? We do not know.

Here the hard evidence ends. His last words were to his brother: "Don't cry. I need all my courage to die at twenty." At his funeral, three thousand people were ready to attack the police. But at the last moment, the general uprising was postponed a few days until what seemed a more auspicious occasion, the funeral of a prominent general appointed by Napoleon. If it was meant for political purposes, Galois's death was in vain.

But his life was the stuff of legend. In 1870, Raspail used the examples of Abel and Galois in a powerful oration in the French Chamber of Deputies that excoriated the established order for its cruelty and indifference to young genius. Yet already in the 1840s mathematicians began to read and study Galois's posthumous writings, including a letter dated on the eve of his duel. This testament became the centerpiece of his legend, as if Galois had feverishly scrawled his great theory on the edge of death. "I have no time," he wrote in the margin, the tragic epitome of doomed genius. Again, the truth is less romantic. He was often too impatient to spell out the tedious details of proofs; he "had no time" for that. Indeed, many of Galois's ideas were already present in his earlier published works (he published five papers in his lifetime) and unpublished papers, though the last letter contains indications of the larger sweep of his ideas.

Galois's work agrees entirely with Abel's results. What is new in Galois is a turn toward abstraction in an essentially modern way, leading to a complete understanding of solvability, which Abel lacked. As he reformulated and extended Abel's work, Galois found it expedient to discuss permutations of the roots of equations not case by case but through a master abstraction that would encompass many permutations at once. Galois was the first to speak of these permutations as instances of a new kind of mathematical object: the *group*. In so doing, he opened the way to the characteristic language and projects of modern mathematics, with all its generality and power. At the same time, however, he opened a crucial gap between the new language and common understanding. That which empowered modern mathematics left many highly educated persons clueless. It remains a great question to what extent this gap can be made intelligible, much less bridged.

# 8    Seeing Symmetries

Once, a mathematician was giving a talk and had just stated a complex theorem bristling with abstract concepts and symbols. Before he could begin to prove it, suddenly someone in the audience blurted out: "Wait! Is that really true?" The speaker paused and drew a small equilateral triangle on the board. He labeled its vertices *A, B,* and *C,* as a schoolchild might. He stared at his triangle for a while, then erased it. "Yes, it *is* true. Let *G* be a group...," going back to his theorem. Even someone comfortable with abstraction felt the need to think about a simple example before moving to the abstract statement.

Likewise, let us seek an intuitive view of the process of solving equations before returning to the story of abstractions. Here, symmetry and permutation are central issues. To show their importance, I will present them as movements in a dance.

Consider first a quadratic equation. In general, it has a pair of roots. To visualize their symmetry, think of two dancers who always return to the same line from which they began and to the original distance between them. They can either stay still or exchange places:

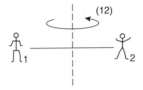

Changing places pivots them halfway around (180°) and corresponds to exchanging the two roots. Changing places twice means a full rotation (360°) that returns them to their original positions.

Let us call this dance $S_2$, because it is symmetric and has two dancers. It consists of two elements: (1), meaning "do nothing," which we will also call "the identity," and (12), meaning "exchange dancers 1 and 2." To describe the dance more fully, we also list the ways the basic steps can be combined. The possible combinations of these steps in sequence can be recorded in, a "Cayley table," introduced by Arthur Cayley in 1854 (table 8.1). We read these steps left to right. For instance, $(12) * (12) = (1)$ means that two exchanges return us to the starting point, the identity. Since the "do-nothing" step changes nothing, $(12) * (1) = (1) * (12) = (12)$. Here, the order in which you do the steps does not matter, so that $S_2$ is "commutative" or *abelian*, to use the modern term that

Table 8.1

| $S_2$ | (1) | (12) |
|---|---|---|
| (1) | (1) | (12) |
| (12) | (12) | (1) |

honors Abel's discovery of commutation as a characteristic of his solvable equations.

At the same time, $S_2$ is *cyclic*, meaning that if the dancers keep repeating one step, (12), over and over, they will generate all the steps of the dance, (1), (12), (1), (12), (1), .... Call this a "2-cycle," since they return to the original pose (1) after two repetitions of (12).

Now $S_2$ could represent the permutations of any two objects, but we are specifically interested in the two roots of a quadratic equation. Since the roots of a quadratic equation are paired by the $\pm$ in its solution, we must seek them *as a pair*. By "completing the square," we find the square root whose two values can switch back and forth in a 2-cycle, just like the two dancers. We have just seen that this cyclical dance is also abelian. Here emerges our basic insight: *Solving an equation corresponds to a certain commutative symmetry*. We will gradually test and refine this insight, which will require subtle qualification.

The analogy between $S_2$ and the process of solving a quadratic equation rests on a parallelism between the steps of a dance and the steps of solving an equation. In each case, there is something invariant, like the central axis around which the two dancers turn. Here we anticipate an important general insight: *Where there is a symmetry, there must be an invariant*. We will now follow this insight in the case of more complicated dances and equations.

As quadratic equations involve two dancers, cubic equations involve three. Think of them arranged at the corners of an equilateral triangle and moving so that they always return to this same triangular formation, though perhaps with different dancers at different vertices of the triangle. As we shall see, there are just six steps they can take: three different exchanges of two dancers, two different rotations involving all three dancers, and the identity. As before, we seek the invariant axis around which the dancers move in each case.

Consider the different ways the dancers can exchange places in pairs. We label the dancers 1, 2, and 3 (perhaps with different costumes). Here is how they are arranged to start, equidistant and viewed from above:

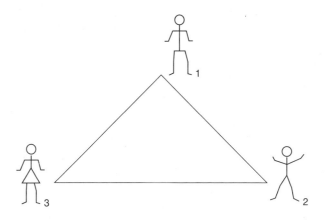

Dancers 2 and 3 can exchange, while dancer 1 stays still, which we call (23):

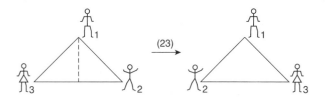

Notice the symmetry axis, which goes from the still dancer to the midpoint of the two exchanging dancers. (If a dancer does not move, we leave its number out of the symbol.)

There are two other possibilities: (12) means that 3 is still, while 1 and 2 change places:

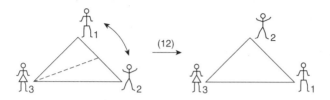

and finally (13) means 2 is still, while 1 and 3 change places:

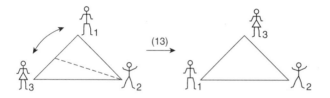

Note that each exchange is an *odd* permutation, since it involves transposing only *one* pair of dancers. (The identity may be considered an even permutation, since it involves zero pairs.)

Since each exchange is an odd permutation, we can put together two of them to make an even permutation. For instance, consider first exchanging (12) and then exchanging (13):

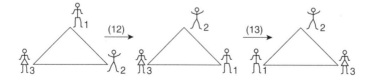

The result of these two sequential exchanges is the same as a rotation (123), meaning that the dancers have rotated by one-third of a complete circle (120°), so that 1 becomes 2 (that is, 2 stands where 1 used to stand), 2 becomes 3, and 3 becomes 1. Looking at them from above, we see:

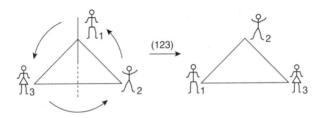

For simplicity, we record the angle of rotations in terms of counterclockwise motion. Note that the rotations involve a single axis standing vertically at the center of the triangle, like a maypole.

In order to obey the rule that the dancers always return to the original triangular formation, they can only rotate by 120° (one-third of a complete circle) or multiples of it. We have just described a single shift of 120° by (123). Likewise, (132) will shift the dancers over by 240°, so that 1 becomes 3, 3 becomes 2, and 2 becomes 1:

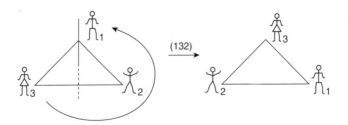

We can build up all the possible rotations that respect the original triangle out of repetitions of the basic rotation of 120°. Thus, two successive rotations of 120° are the same as one rotation of 240°, which we can write as (123) ∗ (123) = (132):

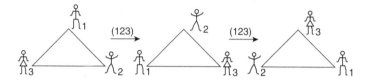

Three successive rotations by 120° add up to a full rotation of 360°, returning us to the identity: (123) ∗ (123) ∗ (123) = (132) ∗ (123) = (1):

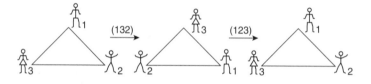

Thus, the rotations of the three dancers are cyclic: the basic rotation (123) repeated over and over gives a "3-cycle": (123), (132), (1), (123), (132), (1), . . . . Notice also that these cyclic rotations by themselves are abelian: whatever the order in which you do them yields the same result. So the dance $S_3$ has six steps: the identity (1), the two rotations (123) and (132), and the three exchanges (12), (13), (23).

$S_3$ is more complex than $S_2$ because $S_3$ is *not* commutative. For instance, if you rotate (123) and then exchange (13), you get (12):

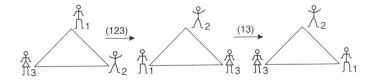

But if you first exchange (13) and then rotate (123), you get (23):

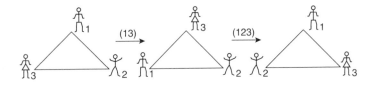

Thus, $S_3$ is nonabelian: When you mix rotations and exchanges, the result is not commutative. Note also that rotations are always even permutations, whereas exchanges are odd. So from the whole nonabelian dance $S_3$, containing both odd and even permutations, we can single out just those that are even, which we will call $A_3$ (for "alternating"), shown shaded in table 8.2, and which are rotations without exchanges. Notice that this set of rotations is "closed" in the sense that these rotations stay within themselves: two sequential rotations always give another rotation, never an exchange. Of course, we must include the identity (1) among

**Table 8.2**

| $S_3$ | (1) | (123) | (132) | (12) | (13) | (23) |
|-------|-----|-------|-------|------|------|------|
| (1)   | (1)   | (123) | (132) | (12)  | (13)  | (23)  |
| (123) | (123) | (132) | (1)   | (23)  | (12)  | (13)  |
| (132) | (132) | (1)   | (123) | (13)  | (32)  | (12)  |
| (12)  | (12)  | (13)  | (23)  | (1)   | (123) | (132) |
| (13)  | (13)  | (23)  | (12)  | (132) | (1)   | (123) |
| (23)  | (23)  | (12)  | (13)  | (123) | (132) | (1)   |

$S_3$ is nonabelian, shown by the lack of symmetry about the main diagonal. For instance, $(123) * (13) = (12)$, but $(13) * (123) = (23)$. The shaded entries show the abelian subgroup $A_3$, which is symmetric about the diagonal.

the rotations, since it corresponds to a rotation by $0°$ (or any multiple of $360°$).

We will call $A_3$ an "invariant subgroup" of $S_3$, meaning that $A_3$ involves some (but not all) of the steps of $S_3$ and is closed and invariant. This term (defined technically in the notes) means that if an invariant subgroup involves one kind of step, such as the rotation (123), it involves all other steps of the same kind—here the rotations (123), (132), and (1). If we look within $A_3$, we find no other such invariant subgroup except the identity itself. For instance, (1) and (12) are a subgroup but not invariant because the other exchanges (13) and (23) are not included.

Here is a symbolic way of writing down this interrelationship of the dances:

$S_3 \triangleright A_3 \triangleright I,$

which simply means: $S_3$ contains the invariant subgroup $A_3$, which in turn contains only the identity, $I = (1)$. The chain goes from the "largest," most inclusive symmetry, to its next largest invariant subsymmetry, then the next largest subsubsymmetry, and so on, ending with the smallest of all, the identity. This nested structure mirrors the solution of a cubic equation. "Completing the cube" corresponds to going from $S_3$ down to $A_3$. Solving a cubic equation corresponds to finding an invariant subsymmetry ($A_3$) within the next larger symmetry of the equation, $S_3$.

Now we take a breath and get ready for four dancers, whose possibilities go far beyond a simple square pattern. If you include all possible steps, the square becomes so tangled that it no longer represents the geometry of the dance. For instance, consider a step like (234): 1 stays still, while 2, 3, and 4 rotate among themselves, not at all in a square. To represent such a step, we need a solid figure. A tetrahedron does nicely: (234) corresponds to dancer 1 standing still, defining an axis around which 2, 3, and 4 rotate by 120°:

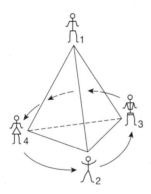

Thus, the full symmetry of a tetrahedron, $S_4$, including both rotations and exchanges, corresponds to the dance of four. In addition, $S_4$ is the rotational symmetry of a cube or

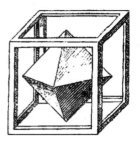

**Figure 8.1**
Johannes Kepler's figure showing an octahedron constructed within a cube by joining the midpoints of its faces, showing that they share the same symmetry (*Harmonices mundi*, 1619).

an octahedron (figure 8.1). These solids have three axes of symmetry:

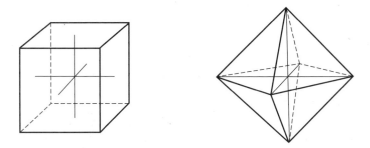

The four diagonals of a cube also correspond to the four dancers or to the four roots of a quartic.

$S_4$ is a complex dance, with $4! = 24$ possible steps. These are so numerous that it is difficult to show the table of $S_4$, which has $24 \times 24 = 576$ entries. Not surprisingly, $S_4$ is nonabelian, for its exchanges and rotations in general are not commutative. The subgroup of even permutations, $A_4$, is nonabelian. Its table has $12 \times 12 = 144$ entries, too many to list here.

Table 8.3

| $V$ | (1) | (12) (34) | (13) (24) | (14) (23) |
|---|---|---|---|---|
| (1) | (1) | (12) (34) | (13) (24) | (14) (23) |
| (12) (34) | (12) (34) | (1) | (14) (23) | (13) (24) |
| (13) (24) | (13) (24) | (14) (23) | (1) | (12) (34) |
| (14) (23) | (14) (23) | (13) (24) | (12) (34) | (1) |

$V$ is an abelian subgroup of $A_4$.

Though $A_4$ is nonabelian, it contains a subsymmetry $V$ that *is* abelian, shown in table 8.3. $V$ is invariant because it includes all the double exchanges, (12)(34), (13)(24), (14)(23), and (1), in which all four dancers participate, but in separate pairs, like the do-si-dos of square dancing. $V$ is abelian (though *not* cyclic) because the double exchanges do not interfere with each other and hence can occur in any order. Having found within the full dance an invariant, abelian subgroup, we now know that the quartic equation, too, can be factored and solved as Ferrari did.

In this progression, something very beautiful has happened. One by one, the Platonic solids have appeared. They have been conjured up by mathematical necessities that are purely matters of permutations, without any mention of geometry. Having encountered the tetrahedron, cube, and octahedron, we might ask, where are the other two, the dodecahedron and icosahedron?

Just as with the preceding cases, we expect that the quintic will be illuminated by the dance of five, $S_5$. From our experience with $S_4$, we fully expect that the dancers' moves will not be adequately represented by a pentagon or any other plane figure, since these cannot represent possibilities such as (12)(345), two dancers exchanging places while the other three rotate among themselves. We must look to solid figures, as before.

*But no regular polyhedron has the symmetry $S_5$.* It would be tempting to draw a connection between the fact that no such polyhedron exists and the unsolvability of the quintic, but such a "deduction" would falsely ascribe a cause and effect relation when all we have is an analogy. There is no intrinsic relation between the properties of three-dimensional space and the solution of equations of the fifth degree. Yet it is amazing that neither the polyhedron of symmetry $S_5$ nor the solution of a quintic equation with the same symmetry exists. If this is a coincidence, it is a beautiful one. There are so few fundamental symmetries ($S_2$, $S_3$, $S_4$, $S_5$) that they necessarily appear in any simple algebraic or geometric situation.

The dance $S_5$ has $120 \times 120 = 14,400$ possible combinations of two steps and is nonabelian. Just the table for its even steps, $A_5$, numbers $60 \times 60 = 3,600$ entries. $A_5$ is also nonabelian: all the exchanges and rotations interfere with each other so that the order of steps matters. At this point, we regain contact with a geometric model, for $A_5$ describes the rotational symmetry of both the icosahedron and the dodecahedron (figure 8.2). Here the final two Platonic solids find their place in the scheme. Their five-fold symmetries correspond to the patterns of five dancers (figure 8.3). As before, let us examine these solids to look for further subsymmetries within them, seeking one that is invariant within the next higher symmetry. If we could find even one, that would correspond to the possibility of solving the quintic equation through taking roots.

In seeking these invariant subgroups, we must consider not only rotations about just one of the axes of the icosahedron,

**Figure 8.2**
Kepler's diagram showing an icosahedron constructed within a dodecahedron by joining the midpoints of its faces, showing that both of these figures share the same symmetry.

for each type of symmetry, but rotations about all of them. There are different sorts of rotations possible about different axes, as box 8.1 shows. If an invariant subgroup includes a certain rotation, it also includes rotations of the same magnitude around any other axis that admits such rotations.

But there is no such invariant subgroup, other than the identity. Box 8.1 shows that the symmetries of the icosahedron are so interconnected that each leads to all the others. Let each pair of dancers make all the possible exchanges. In the process, they will eventually go through all 3,600 combinations of steps. No invariant subgroup of steps remains closed within itself, when it is extended throughout all the possible axes that are equivalent. The beautiful, intricate pattern of $A_5$ is so completely interwoven that it cannot be broken down into separable parts. Thus, there is no invariant subgroup

**Figure 8.3**
Within a dodecahedron, group its twenty vertices into five sets of four equidistant vertices. Connecting each of these five sets yields five intersecting tetrahedra. They show the five-fold symmetry shared by the dodecahedron, icosahedron, and the "dance of five," $A_5$.

smaller than the whole noncommutative dance itself. Therefore the quintic equation cannot, in general, be solved.

What we have called a set of possible dance steps, modern math calls a group. In the abstract sense, a group is a set of elements having a particular structure, defined by four central requirements: The group is *closed* under a certain operation (call it $*$), so that if $a$ and $b$ are elements of the group, $a * b$ is also an element. There is an *identity* element $I$ (so that $a * I = I * a = a$). Each element $a$ has an *inverse* $a^{-1}$

(so that $a^{-1} * a = a * a^{-1} = I$). The *associative law* holds: $a * (b * c) = (a * b) * c$. The earliest examples of groups were permutation symmetries like $S_3$, $S_4$, $S_5$, which is what Galois had in mind when he first used the word "group" in this sense. Here, the elements are permutations, their group operation $*$ is successive permutation, and the identity $I$ is the nonpermutation, (1).

Gradually, the concept of group became more and more general, including not only permutations but any set of elements that would satisfy the basic definition. It was only in 1854 that Cayley set out the basic definition of the theory of groups in this abstract form. He also showed that any finite group is identical in structure to some permutation group. Now free of any visual or intuitive content, the theory found new power in its abstractness. The concept of group thus represents a step in generalization that recalls the step from some particular pair of objects to the number 2. Another such step occurred when Viète went from a particular number like 2 to a coefficient $a$ in a general equation, which might have the value 2 in some case but

---

**Box 8.1**
The symmetry $A_5$ of an icosahedron has no proper invariant subgroup

Consider the different classes of symmetries of an icosahedron, which correspond to the 60 elements of $A_5$. Note first that an icosahedron has 20 faces (each of them an equilateral triangle), 12 vertices (points at which 5 lines meet), and 30 edges (lines connecting two adjacent vertices). There are 6 axes connecting the opposite vertices, each of which allows 4 different "pentagonal" rotations (each a multiple of $\frac{1}{5}$ of a

**Box 8.1** *Continued*

full rotation, since 5 faces surround each vertex in a pentagonal array). These rotations comprise two distinct classes
within $A_5$, each having $6 \times 2 = 12$ elements, corresponding
to odd or even multiples of rotations by $72°$. Then there are
10 axes connecting the midpoints of opposite faces, around
each of which we can make 2 different "triangular" rotations
(each $\frac{1}{3}$ of a full rotation, since the faces are triangles), hence
$10 \times 2 = 20$ elements more. There are 15 axes connecting
the midpoints of the 30 edges; around each of these we can
do an exchange, making 15 elements more. Finally, there is
the identity. So the 60 symmetries have the classes given by
$60 = 12 + 12 + 20 + 15 + 1$.

For instance, consider the "pentagonal" rotations. It turns
out that each of them can be made up of a sequence of
"triangular" rotations,

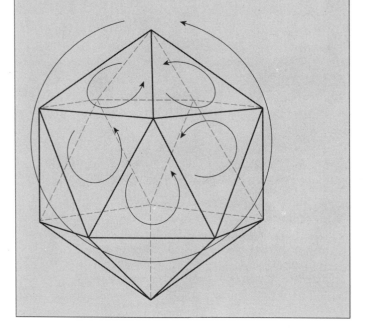

**Box 8.1** *Continued*

so that these two classes of steps are interconnected and are not separate subgroups. Furthermore, each "triangular" rotation can be made up of two exchanges:

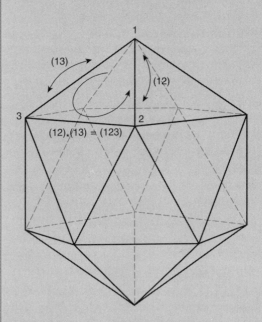

Thus, the possible subsymmetries of $A_5$ are made up of all the possible exchanges, and there is no invariant subgroup that does not include the whole.

These arguments are confirmed by an important general rule, called Lagrange's Theorem: If a symmetry has $n$ elements, any subgroup must have a number of elements that is a divisor of $n$. (A proof of this theorem is given in appendix C.) $S_3$ has 6 elements, while its invariant subgroup $A_3$ has 3, a divisor of 6. Likewise, $S_4$ has 24 elements, while

**Box 8.1**  *Continued*

$A_4$ has 12 and $V$ has 4 elements, both divisors of 24. (Remember that the identity is always counted as an element of the subgroup.) But $S_5$ has 120 elements and $A_5$ has 60, a divisor of 120. Now any subgroup of $A_5$ must be a divisor of 60. We have found that the possible invariant subgroups are given by the list $60 = 12 + 12 + 20 + 15 + 1$. Now any subgroup would both have to be a divisor of 60 *and* also be composed of either $12 + 1 = 13$, $20 + 1 = 21$, or $15 + 1 = 16$ elements. But neither 13, 21, nor 16 (nor any sum of them) is a divisor of 60. So the only invariant subgroups of $A_5$ are "improper": either all of $A_5$, all 60 elements, or just the identity, because 60 and 1 are the only divisors of 60 that are possible. Thus, $A_5$ has no proper invariant subgroup.

in general has many possible values. The concept of group allows us to contemplate pure relationships without the distraction of considering the things related. Moreover, there are many possible groups, each with its own peculiarities, each capable of being realized by many possible sets of elements. Some groups are "trivial," as mathematicians put it, having one sole member, the identity, $I$. As we have seen with $S_5$, each group has its own character and peculiar beauty, sometimes complex, even bizarre.

The concepts of group theory parallel every step of our dance analogy. The symmetries we discussed—$S_2$, $S_3$, $S_4$, $S_5$— are each groups with different numbers of elements. Each of these groups contains subgroups, as $S_4$ contains the subgroup $A_4$, expressing the interrelation of the subsymmetries within the next larger symmetry. Groups can be abelian (commutative) or nonabelian. What we called invariant subgroups are usually called "normal" (see notes to this chapter). The nesting of symmetry and subsymmetry is formalized as a "solvable chain," written (in the case of the cubic equation) as

$$S_3 \triangleright A_3 \triangleright I,$$

where $S_3 \triangleright A_3$ means: "$S_3$ contains $A_3$ as a normal subgroup." This was Galois's great advance over Abel, specifying exactly which equations are solvable and which not. By 1889, Camille Jordan and Otto Hölder expressed this in a form that highlights commutativity: An equation is solvable if the "quotient group" of each successive pair of links in the chain is abelian. That is, the chain shows how the group is made up of "building blocks," namely invariant subgroups nested and linked, in which each successive link is as large as possible. To be solvable, each successive link must fit into the next in an abelian way. Here, the criterion of commutativity is subtle, for it is *not* necessary that the *links* be abelian, only their "linkage." The concept of "quotient group" (defined and illustrated on pp. 176–177, 193–194) gives precise form to this.

A group having no proper invariant subgroups is called "simple," though (like the icosahedron and its symmetry group, $A_5$) intuitively it may be formidably complex. Even so, $A_5$ turns out to be the smallest such nonabelian group, in the sense of having the smallest number of elements. A great tour de force of twentieth-century mathematics was the exhaustive enumeration of finite simple groups, starting with $A_5$ and including the so-called "monster," which corresponds to a certain symmetry of a solid in a space of 196,883 dimensions. Alas for intuition, with nothing to visualize or grasp. But the abstract theory stands ready to help when intuition is overwhelmed and clueless. Here the question of commutativity can be the guiding thread.

# 9          The Order of Things

Abel first noticed the commutativity in his special equations, so it is only fitting that this property is named for him. Though Galois immensely extended Abel's work, he did not emphasize the theme of noncommutativity, nor did those who later systematized his ideas, such as Jordan and Hölder. Doubtless their attention was on precise mathematical formulation rather than philosophical implications. Nevertheless, their insight entered deep into the mainstream of mathematics, largely through the influence of the theory of equations they shaped. So intrinsic is the thread of noncommutativity that modern treatments take it for granted. That being the case, it is important to reconsider its import and seek ways to express it.

For the first generations that followed Abel and Galois, this new theory was still rich and strange. In retrospect, Galois's radical viewpoint won out over the older approach because it gave a complete understanding of solvability, where Abel's results were only partial. Yet it is worth pausing to consider the confrontation between them and how it played out, for it reflects deep currents in the development of mathematics.

At the beginning of the nineteenth century, algebra was still regarded as an offspring of arithmetic, closely tied to its fundamental operations. Gauss considered $a \times b = b \times a$ the first "principal truth about multiplication." In 1830, George Peacock was the first to try to give algebra "the character of a demonstrative science." He distinguished between "arithmetical" algebra, whose elements are numbers and whose operations are those of arithmetic, and "symbolical" algebra, "a science, which regards the combinations of signs and symbols *only* according to determinate laws, which are altogether independent of the specific values of the symbols themselves." This distinction is not far from Viète's contrast between the logic of numbers and of species. It emphasizes the abstract quality of the symbols, opening the door to new possibilities, though Peacock thought of the symbols fundamentally as generalizations of ordinary numbers.

Peacock also set forth what he called a "principle of the permanence of equivalent forms," which essentially means that, though the symbols may be general, the laws of algebra should always be the same as those of arithmetic. For instance, Cauchy wrote that "if $a$ and $b$ be whole numbers, it may be proved that $ab$ is identical with $ba$; *therefore*, $ab$ is identical with $ba$, whatever $a$ and $b$ may denote, and whatever may be the interpretation of the operation which connects them." There is no hint that these symbols could ever escape from the laws of ordinary numbers.

Peacock's contemporaries gradually threw off these restrictions. In 1830, Augustus De Morgan wrote that "no word or sign of arithmetic or algebra has one atom of meaning," so that their interpretations should be open and arbitrary, not restricted to ordinary numbers and magnitudes. By 1840, Duncan Gregory had added that these symbols could represent operations, not just numbers. Building on this, in 1847 George Boole used algebraic symbols to represent members of a "set," where the whole universe of discourse is called

1 and the empty set 0. In "Boolean algebra," addition may mean uniting two sets, multiplication intersecting them. Let 1 mean the totality of humans, $x$ all Americans, and $y$ all women. Then $(1 - x)$ means all non-Americans, $(x + y)$ all Americans and all women, $xy$ all American women. Not only did Boole show the close relation of logic to mathematics, he also emphasized that the form, rather than the content, of the symbols is crucial. For this reason, Bertrand Russell credits Boole with having discovered pure mathematics, "the greatest discovery of the nineteenth century."

The way now stood open for algebras that would break decisively with the rules of ordinary arithmetic. The first was found by an Irishman, William Rowan Hamilton. He was an early admirer of Abel's proof and wrote an extensive account of it that helped other mathematicians at a time when they found it puzzling in the extreme. As mentioned earlier, Jerrard thought he had found a solution to the quintic in 1835, which stimulated Hamilton to study Abel's proof carefully. Hamilton tried tactfully to persuade Jerrard that his proof was mistaken, but Jerrard died unconvinced, still defending his proof in 1858, which shows how long it took for Abel's ideas to be widely accepted. In Hungary during the same period, the brilliant János Bolyai, co-discoverer of the first non-Euclidean geometry, also tried to solve the quintic, evidently unaware of Abel and Ruffini.

By emphasizing the power and generality of Abel's insights, Hamilton showed his contemporaries how important and beautiful Abel's theory really was. In so doing, he helped it to live and to flourish in other minds, so that it would not be neglected because of its novelty and difficulty. In France, Joseph Liouville and Camille Jordan performed an even more important service for Galois, whose fragmentary and abbreviated writings needed extensive amplification before their import could emerge.

Such generous admiration brought a rich reward. Beginning in 1833, Hamilton sought a new kind of number that would extend the idea of complex numbers. He realized that a complex number like $a + bi$ (where $i = \sqrt{-1}$) could be thought of as an ordered pair composed of its real and its imaginary parts, $(a, b)$. As noted earlier, such an ordered pair could be represented as a point in a plane (see figure 3.2). Hamilton then sought to generalize this idea, from the plane to three dimensions, from an ordered pair $a + bi$ to an ordered triple like $a + bi + cj$, where $j$ would generalize $i$ so that $i^2 = j^2 = -1$.

For ten years, Hamilton tried to make these ordered triples work. He could understand how they should add (term by term, grouped according to common factors of 1 or $i$ or $j$), but not how they should multiply. In 1843, the answer came to him suddenly, as he and his wife were walking along the Royal Canal in Dublin. His difficulties could not be solved with triples, but could be if he took quadruples, $a + bi + cj + dk$, where $i^2 = j^2 = k^2 = -1$. To make multiplication work for these "quaternions" (as he called them), he had to make this operation noncommutative: $ij = k$ but $ji = -k$, and likewise $jk = i = -kj$ and $ki = j = -ik$. Since $ij = -ji$, reversing the order of multiplication reverses the sign of the product, so that one might say that these quantities *anticommute*. To mark this special moment, Hamilton took out his pocket knife and carved the equation $i^2 = j^2 = k^2 = ijk$ (since $ijk = kk = -1$) into a stone of the Brougham Bridge.

The carving has long since worn away, but the strange beauty of noncommutative numbers lives on. Hamilton spent the rest of his life developing the new algebra of quaternions, which he described as "a curious offspring of a quaternion of parents, say of geometry, algebra, metaphysics, and poetry," reflecting his devotion to philosophy and poetry (he wrote poetry enthusiastically, if not well, and was a friend of Samuel Taylor Coleridge). He gradually developed quaternions, leading to the modern notion of vectors (magnitudes with direction) and scalars (pure magnitudes with no direction).

Hamilton noted that there were two kinds of multiplication that emerged. One he called the "scalar product," which takes two vectors, is commutative, and produces a scalar as a result. The name has stuck. The other (not named) was not commutative and produced a vector as a result. Hamilton believed that his quaternions promised a new view of the cosmos, and a number of British mathematicians also took them as the key to mathematics and mathematical physics. The great Scottish physicist James Clerk Maxwell, though interested in quaternions, finally did not join the true believers. The difficulty was that, just as complex numbers imply a two-dimensional plane, the four-component quaternion seemed to require a four-dimensional "space." At that time, there was no hint of what that could possibly be.

As if responding to the same unspoken question, Hermann Grassmann, a secondary-school teacher in Germany who became a specialist in Sanskrit, independently developed an algebra based on a "space" of $n$ dimensions, not merely three. In 1844, Grassmann published his "Theory of Extension" (*Ausdehnungslehre*), which distinguished "inner" products, commutative like Hamilton's "scalar product," from "outer" products that were not. As he studied the possibilities, Grassmann proved that "for products of two factors there are... only two types of linear product, namely the one whose system of determining equations has the form $e_i e_j + e_j e_i = 0$ and the one for which it has the form $e_i e_j = e_j e_i$, in which the indices $i$, $j$ can go from 1 to $n$," the number of dimensions in the "space." That is, either the factors must commute ($e_i e_j = +e_j e_i$) or they must anticommute ($e_i e_j = -e_j e_i$). Since they are not restricted to any specific number of dimensions, Grassmann's algebras emphasized the generality of commuting versus anticommuting.

Grassmann's work was only slowly assimilated because it was expressed in unconventional notation. Gradually, others began to express similar ideas in their own ways. Indeed, Oliver Heaviside and Josiah Willard Gibbs had independently found similar results. Gibbs reformulated these ideas in the language of vectors that is now used universally. Following Gibbs, Grassmann's "outer" product became what we call the "cross" or "vector" product, in which $\vec{A} \times \vec{B} = -\vec{B} \times \vec{A}$, as with Hamilton's quaternion multiplication.

This noncommutative algebra is the basis of the modern understanding of physics, whether Newton's mechanics or Maxwell's theory of electricity and magnetism. Found in every textbook, it is routinely taught without much attention to its striking noncommutativity. It is also central to matrix algebra, which was developed beginning about 1858 by two inseparable English friends, the "invariant twins" Arthur Cayley and James Joseph Sylvester. The idea of matrices emerges from looking at the coefficients of simultaneous equations and arranging them in an array. Once again, the problem of solving equations stimulated research in a new direction, for matrix algebra allows great simplification of the process of solution (box 9.1). Soon it was realized that matrices, too, have noncommutative multiplication (box 9.2) and can be used to represent the algebras of Hamilton and Grassmann. Matrices show that such algebras do not necessarily require "complex" or exotic kinds of number, for matrices with real values (but in an ordered array) can realize every algebraic requirement of complex numbers or of quaternions. Indeed, matrices are perfect vehicles for expressing the

**Box 9.1**

Two successive linear transformations, compared with their matrix form

Consider a transformation (such as a rotation of coordinates) that takes the variables $x$ and $y$ and defines new variables $x'$ and $y'$:

$$a_{11}x + a_{12}y = x'$$
$$a_{21}x + a_{22}y = y'.$$

Consider a second such transformation that expresses $x'$, $y'$ in terms of $x''$, $y''$:

$$b_{11}x' + b_{12}y' = x''$$
$$b_{21}x' + b_{22}y' = y''.$$

Expressing $x''$, $y''$ in terms of $x$, $y$ involves solving and re-substituting these two transformations, but can be expressed much more simply if we write them in matrix form:

$$\begin{pmatrix} b_{11} & b_{12} \\ b_{21} & b_{22} \end{pmatrix} \begin{pmatrix} a_{11} & a_{12} \\ a_{21} & a_{22} \end{pmatrix} \begin{pmatrix} x \\ y \end{pmatrix} = \begin{pmatrix} x'' \\ y'' \end{pmatrix}$$

The two matrices can be multiplied using the rule that each row of the first matrix be multiplied term-by-term by the first column of the second matrix, and the sums taken of those products:

$$\begin{pmatrix} b_{11} & b_{12} \\ b_{21} & b_{22} \end{pmatrix} \begin{pmatrix} a_{11} & a_{12} \\ a_{21} & a_{22} \end{pmatrix} = \begin{pmatrix} a_{11}b_{11} + a_{21}b_{12} & a_{12}b_{11} + a_{22}b_{12} \\ a_{11}b_{21} + a_{21}b_{22} & a_{12}b_{21} + a_{22}b_{22} \end{pmatrix}$$

This notation greatly simplifies dealing with systems of equations.

operations of groups. Grassmann's generalization of the possibilities of algebra opened an ever-widening field. Cayley discovered an eight-dimensional generalization of quaternions, now called octonions or Cayley numbers. Sylvester called himself "the new Adam," giving names (not all of

**Box 9.2**
Matrix multiplication is not commutative in general

Consider, for instance, the following noncommuting matrices:

$$\begin{pmatrix} 0 & 1 \\ 1 & 0 \end{pmatrix}\begin{pmatrix} 1 & 0 \\ 0 & -1 \end{pmatrix} = \begin{pmatrix} 0+0 & 0+(-1) \\ 1+0 & 0+0 \end{pmatrix} = \begin{pmatrix} 0 & -1 \\ 1 & 0 \end{pmatrix},$$

which is not equal to the product taken in the opposite order:

$$\begin{pmatrix} 1 & 0 \\ 0 & -1 \end{pmatrix}\begin{pmatrix} 0 & 1 \\ 1 & 0 \end{pmatrix} = \begin{pmatrix} 0+0 & 1+0 \\ 0-1 & 0+0 \end{pmatrix} = \begin{pmatrix} 0 & 1 \\ -1 & 0 \end{pmatrix}.$$

which stuck) to a whole new universe of algebraic beings: invariants, discriminants, zetaic multipliers, allotrious factors, among many others. By 1860, Benjamin Peirce had enumerated 162 different algebras. (His remarkable son C. S. Peirce carried forward Boole's logical work as a philosopher, besides being a chemist and astronomer.)

The historical unfolding of the notion of noncommutativity implies larger questions: What is its significance? Why should we care about it? In 1872, not long before Cayley set out group theory in full abstraction, the great mathematician Felix Klein gave a lecture at Erlangen, in which he proposed that group theory be considered the heart of mathematics, more fundamental in its import than geometry or algebra alone. He argued that each geometrical figure should be considered more in terms of its fundamental symmetry—specified by the group governing it—than by any figure or equation in itself. In Klein's view, considering ever more general groups would be the royal road to understanding ever more complex spaces, reaching past three dimensions to manifolds of higher dimensions.

The "Erlangen Program" was the clarion call of a new abstract vision, going past the traditional starting points of mathematics to reveal the deepest, most general concepts underneath. It was an influential rallying cry that still resounds in contemporary mathematics. As part of his program, Klein went back to reconsider the Platonic solids in terms of group theory. Where the Greeks considered these solids as emerging from the necessity of fitting together regular polyhedra in three-dimensional space, Klein saw them as incarnations of fundamental groups. Thus, an equilateral triangle should be considered most of all in terms of its symmetry group $S_3$, the rotations and reflections that leave it invariant. Similarly, the Platonic solids are incarnations of more complex rotational symmetries: the tetrahedron of $A_4$, the cube and octahedron of $S_4$, the dodecahedron and icosahedron of $A_5$. Indeed, Klein's presentation lies behind the dance analogy used in the preceding chapter.

In fact, noncommutativity is everywhere. In general, rotations in three-dimensional space do not commute. Imagine leaving Santa Fe for two journeys: (a) going directly west for 1,000 km; (b) going directly north for 1,000 km. If you do these journeys in the order first (a), then (b), you will arrive at a certain spot. But if you do them in the reverse order, (b) then (a), you end up somewhere else, about 123 km away. The order matters since you are traveling on a globe, not a flat plane.

Klein considered this peculiar noncommutativity of spatial rotations to be the hallmark of space itself, understood as a manifestation of its underlying symmetry of rotations. He classified geometries by the algebraic properties that remain invariant under a particular group of transformations, as Euclidean areas and lengths do when moved in the plane. This was one of many instances that moved Klein to deem group theory the master key that would unlock the secrets of both algebra and geometry by uncovering what underlies both of them.

Long before, Descartes had noticed that the roots of equations could be increased or decreased, scaled up or down by a multiplicative factor, while leaving invariant their essential constellation: the number of the roots and their relative spacing. Looking back from the perspective of the twentieth century, the eminent mathematician Hermann Weyl took this as a kind of "relativity" that emerged long before Albert Einstein's theories. In 1927, Weyl noted that "it was a lucky chance for the development of mathematics that the relativity problem was first tackled, not for the continuous point space, but for a system consisting of a finite number of distinct objects, namely the system of the roots of an algebraic equation with rational coefficients (Galois theory)." Weyl even wrote that, because of its novelty and profundity, Galois's final letter was "perhaps the most substantial piece of writing in the whole history of mankind." The original development of group theory in this discrete, algebraic context prepared the way for Einstein and Hermann Minkowski's understanding of relativity in terms of groups of continuous motions that leave the speed of light invariant. Here again, the symmetry between all uniformly moving frames of reference points to an invariant.

As the comparison of the regular solids and the solvable equations shows, algebra felt the impact of noncommutativity before geometry, despite such simple cases as the journeys around the Earth just described. Perhaps this is simply because the order in which things are done far more naturally applies to algebra, in which symbols and operations are read in a certain order, than to geometry, which seems to rely on timeless diagrams. Plato considered the eternal changelessness of mathematics a hallmark of its closeness to absolute reality, to pure being beyond existence and becoming. It was, then, a shock to realize that the order of things mattered.

The question of order is crucial for modern physics. The heart of thermodynamics is its Second Law, a statement that natural processes are in general irreversible. Heat flows from hotter to colder bodies, not vice versa, unless we expend work and also expel heat into the rest of the universe. Struck by its sweep and scope, Max Planck emphasized irreversibility as he hesitatingly introduced the quantum into physics. On the human scale, irreversibility means that every action, once done, can never be undone. It gives dignity to human choice by emphasizing its irrevocable consequences. Time is no longer the perpetually reversible stream of Newtonian or Maxwellian dynamics. Instead, irreversible time is the dark river of history, flowing ever onward.

Ironically, Planck himself tried to reverse the flow of events and undo the shock of the quantum, or at least tame it into a more classical form. Ruefully, he recognized that here too irreversibility held sway. As quantum theory took shape, it revealed noncommutativity at its very core, the child of irreversible thermodynamics married to reversible mechanics. Formally, Max Born and Pascual Jordan showed that its crucial realization was that physical quantities, when expressed in algebra, did not necessarily commute. They described the position of a particle $q$ and its momentum $p$ as operators obey the noncommuting relation $pq - qp = ih/2\pi$, where $h$ is Planck's constant ($h = 6.6 \times 10^{-27}$ erg-seconds) and $i = \sqrt{-1}$. Since $h$ is extremely small by human standards, this noncommutativity usually manifests itself only on the atomic scale and is the direct source of the famous Heisenberg uncertainty principle. As I have discussed in my book *Seeing*

*Double*, these are the distinctive hallmarks of the utter indistinguishability of quanta. Noncommutativity continues to govern the deepest level of physical theory, whether the quantum theory of fields at space-time points or alternative string theories that use not points but tiny lengths or membranes. For all of them, noncommutativity is basic and promises to be so in the future.

Initially, noncommutativity in quantum theory stood for the irreversibility of measurement in general. The theory of relativity also relies on noncommutativity to express another fundamental principle: Effects do not follow instantaneously from causes but are limited in their speed by the velocity of light. Here group theory expresses beautifully the symmetries of space-time. If events are sufficiently separated in space and time that no light signal can connect them, then all physical quantities of the events must commute, expressing their independence. Conversely, within the realm of causal influence traveling at or less than the speed of light, physical quantities in general may not commute, expressing their causal relation. Any change in this principle would require overturning this fundamental principle of causality, which has so far been confirmed in myriad ways. Commutativity versus noncommutativity lies at the heart of causality itself.

Going from these most general principles to more specific cases, noncommutativity is crucial to the development of the "standard theory" that has very successfully drawn the known elementary particles into a largely unified scheme. This theory rests on what are called nonabelian gauge fields. Without even exploring what such fields might be in detail, note that their very name announces that they rely on a special sort of noncommutativity in the symmetries of fundamental particles. Such nonabelian theories promise to be the inescapable vehicles of future physical theory, if only as a point of departure. They also are the starting points of attempts to unite quantum theory with Einstein's general theory of relativity, which is a geometrical theory of gravitation as the curvature of space-time. These directions in physics have in turn stimulated further developments in mathematics, notably the theory of "noncommutative geometry." Klein's conjecture continues to be deeply provocative: nonabelian groups may give deep clues to the nature of space and time. As noted in the last chapter, the rotational invariance of three-dimensional space leads to the five Platonic solids. Perhaps such considerations of groups may someday explain why we perceive three dimensions of space and only one of time. The search has only just begun.

# 10    Solving the Unsolvable

Abel's work opened the way to Galois's explanation of the failure to solve all equations in radicals. Indeed, radicals had posed dilemmas since the beginning. The discovery of irrational quantities cast into doubt the nature of mathematics and, with it, the possibility of definitive human knowledge. In Greek, *krisis* means a trial, a decisive act of distinction, separation, judgment. The impossibility of making the irrational commensurate with the rational tested and formed Greek mathematics and philosophy.

Much later, Abel's "impossibility proof" was a watershed for modern mathematics. It had the dimensions of a crisis, though its significance has not been fully acknowledged or assessed. To do so requires a special effort of imagination that would allow us to return to a moment that has been forgotten, the moment in which Abel confronted what he had proved.

As discussed earlier, it is not clear whether the ancients regarded the discovery of the irrational as a disaster or a miracle. It was ominous; one might drown in such an unnameable sea, beyond the security of the counting numbers. As Euclid showed at the end of his Book X, the irrationals lead to an infinite vista, which could inspire horror or wonder. Euclid tried to keep them rigorously separate from the integers, if only to avoid paradox and confusion. Yet what the ancients separated, the moderns united. Gradually but inexorably, the development of algebra led to the term "number" being applied to irrationals as well as integers, for $x = \sqrt{2}$ seems no less valid a solution of the equation $x^2 = 2$ than $y = 2$ is a solution of $y^2 = 4$.

Once symbolic algebra was established, the crisis of unsolvability was inevitable. Abel's proof showed that radicals are not sufficient, just as the ancient proof showed that rationals are not enough. Since solutions of all equations are known to exist by the Fundamental Theorem of Algebra, they must be a new kind of magnitude. The name "algebraic numbers" came to signify all the solutions of algebraic equations whose coefficients are integers. As with the irrationals, giving them a name seemed tantamount to dispelling the sense of crisis or paradox. For mathematicians, new solutions of the quintic eclipsed the question of solution in radicals, for they could simply annex the algebraic numbers to those expressible in radicals. To give a name to those algebraic numbers that are *not* expressible in radicals, we might call them "ultraradical numbers."

Now we can complete the odyssey of the quintic. It turns out that their solutions can be expressed as the so-called elliptic modular functions (which are sums of powers times trigonometric functions). These functions are based on the "theta functions" that Abel and Jacobi began to study in 1827, though it was not until 1858 that Charles Hermite, Leopold Kronecker, and Fernando Brioschi independently showed their relation to the quintic. By 1870, Jordan had proved that any polynomial equation, of whatever degree, could be solved by using these generalized functions.

During the same period, people started building machines that could find approximate solutions. About 1642, Blaise Pascal had the first vision of machines that could compute. In 1840, Leon Lalanne devised a mechanical computer that could approximate the solutions of equations up to the seventh degree. In 1895, Leonardo Torres Quevedo developed an elegant machine that could calculate logarithms needed to solve difficult algebraic problems (figure 10.1). At Bell Laboratories in 1937, scientists developed a larger but less elegant-looking machine, the Isograph, which could deal with equations up to the fifteenth degree (figure 10.2). By the end of the twentieth century, far more powerful programs were available to users of personal computers.

These developments simply sidestepped the problem of solution in radicals. But what, then, happened to the "crisis"? Was it ever resolved? Or is it a case of history being written by the victors, the moderns who successfully bypassed and set

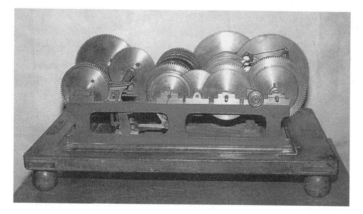

**Figure 10.1**
A machine built in 1895 by Leonardo Torres Quevedo to calculate logarithms.

**Figure 10.2**
The "Isograph," built in 1938 at Bell Laboratories to solve polynomial equations up to the fifteenth degree.

aside the older scruples? Perhaps they were right not to look back; they were so busy forging ahead that they had little time for retrospection. Yet surely there remains valuable insight to be gained by reflecting on what each successive crisis really meant.

Each crisis opened a new insight into the infinite, in different contexts. First, between arithmetic and geometry: the Greek discovery of incommensurability implied that expressing an irrational requires an infinite number of digits. Next, between geometry and algebra: to express the area of an oval requires an infinite number of algebraic terms, as Newton showed. Finally, Abel's proof, within algebra itself: solving a finite equation requires an infinite number of terms.

Even solving a cubic equation with real roots requires that we take the cube root of an arbitrary complex number. As complex numbers were accepted, taking their roots became imperative. The problem was resolved by Abraham De Moivre in 1707 and given its modern form by Euler in 1748. We must start by expressing each complex number in terms of a magnitude and an angle, understood as polar coordinates in a plane. De Moivre's formula then gives a beautiful expression for the $n$th power of the complex number (where $n$ could be any integer or fraction) in terms of sine and cosine functions (box 10.1).

Most calculators provide sines or cosines by pressing a button. Geometrically, these functions are ratios of the sides and hypotenuse of a right triangle. Yet algebraically sines and cosines are not finite polynomials, but infinite series. Newton had already discerned this algebraic infinitude in studying

---

**Box 10.1**
De Moivre's formula

Express the complex number $z$ in terms of its absolute value $r$ and its argument $\theta$, $z = r(\cos\theta + i\sin\theta)$. In modern notation, De Moivre proved that

$$z^n = r^n(\cos n\theta + i\sin n\theta).$$

To show this, consider the two "Taylor series" $\sin x = x - \frac{x^3}{3!} + \frac{x^5}{5!} - \frac{x^7}{7!} + \cdots$ and $\cos x = 1 - \frac{x^2}{2!} + \frac{x^4}{4!} - \frac{x^6}{6!} + \cdots$. If we write $z = re^{i\theta}$, where $i^2 = -1$ and $e = 2.718\ldots$ is the basis of natural logarithms, then $z^n = r^n e^{in\theta}$. Using the Taylor series for $e^{in\theta} = 1 + in\theta + \frac{i^2 n^2 \theta^2}{2!} + \cdots$, De Moivre's formula follows if we group the real and imaginary terms separately:

$$z^n = r^n e^{in\theta} = r^n\left[\left(1 - \frac{n^2\theta^2}{2!} + \cdots\right) + i\left(n\theta - \frac{n^3\theta^3}{3!} + \cdots\right)\right].$$

If we set $n\theta = \pi$ and $r = 1$, then $e^{i\pi} = -1$ (since $\sin\pi = 0$ and $\cos\pi = -1$) or $e^{i\pi} + 1 = 0$, Euler's beautiful formula.

the areas of ovals. For him, this implied that geometry can express in a few lines an infinite series of algebraic terms. Thus, Newton was able to write the sine and cosine as $\sin x = x - \frac{x^3}{3!} + \frac{x^5}{5!} - \frac{x^7}{7!} + \cdots$ and $\cos x = 1 - \frac{x^2}{2!} + \frac{x^4}{4!} - \frac{x^6}{6!} + \cdots$. Using such series, calculators sum a few terms to provide a value of required accuracy. As box 10.1 shows, these series lead to the formula $e^{i\pi} + 1 = 0$, one of Euler's most beautiful thoughts, uniting $e, i, \pi, 1$, and $0$ in one pregnant expression.

Though it was long suspected, finally in 1873 Hermite showed rigorously that $e$ was transcendental, not the solution of any algebraic equation of finite degree, and in 1882 Ferdinand Lindemann did the same for $\pi$. These proofs ended the dreams of the circle-squarers, whose hopes rested on the possibility of an algebraic formulation of $\pi$. One might have thought that such transcendentals were rare and exotic, but $e$ and $\pi$ were only the beginning. In 1874, Georg Cantor showed that there are uncountably many transcendentals, which are far more "dense" in the real line than the countable infinitude of integers or of rational numbers. He also showed that algebraic numbers are countable.

To do these later developments justice would require another book and would take us away from Abel, to whom we now return one final time. His discovery showed that there is an intermediate class between algebraic irrationalities (such as the radicals earlier expected to solve the quintic) and the full generality of transcendental numbers. The ultraradical solutions of the quintic equation, in general, are neither the one nor the other; they are algebraic numbers, but they are not composed of radicals. In this, they are a kind of amphibian between land (the radicals) and the vast transcendental sea. Here, they are only one among innumerable kinds of irrational numbers, whose infinitude was already implicit in Euclid's Book X.

Standing on the edge of this ocean, Abel gazed at the dawn of a new mathematics, which he began to explore in the few years remaining him. Great as it was, his proof of the unsolvability of the quintic came during his school years. It would be unjust to represent it as his only or greatest accomplishment in mathematics. He went on to do profound work on transcendental functions, infinite series, and integrals that generalize trigonometric functions, the "Abelian integrals" and "Abelian Addition Theorem" that deeply influenced those who came after him and that many mathematicians call his greatest work. He also took Gauss's seminal insights even further than the master. Where the young Gauss had divided a circle into seventeen parts to construct a regular seventeen-sided polygon, Abel was able to make a similar division of a more complex curve, the lemniscate.

Nevertheless, Abel called the solvability of equations "my favorite subject." Though not much given to speculating about the significance of what he had done, in the introduction of his final paper on the solution of equations (written in 1828 but published only posthumously in 1839) Abel articulated a kind of credo: "One must give to a problem a form such that it is always possible to solve it, which one can always do with any problem. Instead of looking for a relation that one does not know exists or not, one must ask if such a relation is really possible." He had shown that solvability in radicals is a condition that need not always exist. By transcending this limitation, Abel solved the unsolvable. As if taking up Viète's bold claim "to leave no problem unsolved," Abel recorded the power of the infinite extension of finite algebra. The end of the old assumption that all equations have a finite solution revealed a new mathematics of infinite series and noncommutativity.

There is another document that records something of
Abel's inner feelings about his work and life, but it is cryptic
and private. Written in 1826, during his work on the lemniscate, Abel's doodlings in one of his Paris notebooks (figure 10.3) combine mathematics and reverie in so unguarded

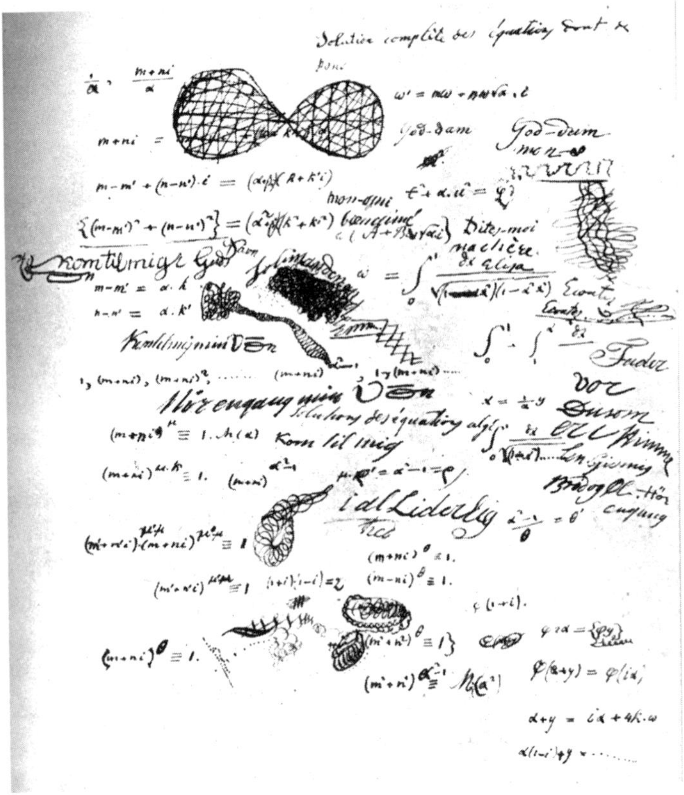

**Figure 10.3**
A page from Abel's Paris notebook (1826). The large lemniscate ($\infty$) is at
the top of the page; the passages discussed in the text are to the right and
under it.

a way that he probably would have been amused if not mortified to learn that it would become a treasure of the National Library in Oslo. This page was doubtless meant for no other eyes than his own.

Abel writes in French and Norwegian, interspersed with equations and drawings, featuring a striking drawing of a lemniscate, a large figure ∞. Abel has cross-hatched this curve and repeatedly outlined it. At the top of this page, Abel writes "complete solution to the equations in which ...," breaking off into equations and intimate thoughts: "My friend ... beloved ... Come to me in God's name ...." In the midst of this, an ironic prayer: "Our Father who art in Heaven, give me bread and beer. Listen for once." Then, in French, "speak to me, my dear," her name oddly entwined in an elliptic integral,

$$w = \int_0^1 \frac{dx \quad Elisa}{\sqrt{(1 - x^2)(1 - \alpha^2 x^2)}}.$$

Then: "Listen ... Listen ... Come to me, my friend ... not, for once, my solutions to algebraic equations ... come to me ... in all your lewdness."

No page more daringly suggests the mysterious intersection of desire and mathematics. Who is Elisa? An unknown beloved, one of the tempting Parisians whose beauty he mentioned in a letter, a remembered or even an imagined woman? Why does he scrawl the name of "Soliman den Anden," the Ottoman emperor Sulieman II who also appears on another page? What of the neat equations interwoven among these dreamy fragments? And what of the words near the top of the page: "Goddamn ... Goddamn my ∞." What is the tone, the felt meaning of this outburst? Is it humor, vexation, wonder?

We will never know. What remains is equations and ∞.

# Appendix A
## Abel's 1824 Paper

*Memoir on algebraic equations, in which is demonstrated the impossibility of solving the general equation of the fifth degree (1824)*

Geometers have been much concerned with the general solution of equations of the fifth degree and many of them have sought to prove their impossibility, but (if I am not mistaken) none has succeeded until now. Therefore I hope that geometers will receive kindly this memoir, which aims to fill this gap in the theory of algebraic equations.

Let

$$y^5 - ay^4 + by^3 - cy^2 + dy - e = 0 \qquad [A1]$$

be the general equation of the fifth degree and let us suppose that it is solvable algebraically, that is, one can express $y$ by a function formed by radicals of the quantities $a$, $b$, $c$, $d$, and $e$.

*[The commentary is italics in brackets []; equation numbers have been added also in brackets. Note Abel's notation for the coefficients, which does not follow the conventions of our text. He has chosen the signs of [A1] so that the coefficient a is the sum of the roots, b the sum of their products, c the sum of their triples, etc., according to Girard's identities (box 4.1).]*

It is clear that in this case we can express $y$ in the form

$$y = p + p_1 R^{\frac{1}{m}} + p_2 R^{\frac{2}{m}} + \cdots + p_{m-1} R^{\frac{m-1}{m}}, \qquad [A2]$$

$m$ being a prime number and $R$, $p$, $p_1$, $p_2$, etc. functions of the same form as $y$, and so on until we come to rational functions of the quantities $a$, $b$, $c$, $d$, and $e$.

*[For Abel's argument for the generality of [A2], which he considered Step I of the larger proof, see appendix B. Essentially, he shows that a finite sum of radicals, however they are nested, can always be expressed in the form [A2], including putting all the terms on a common denominator, so that eventually one reaches only rational functions inside the innermost radicals. Abel then uses [A2] to frame a reductio ad absurdum: he begins by assuming that y can be written as a finite series of algebraic terms, in which R, p, $p_1$, $p_2$, ... are algebraic functions of the coefficients a, b, c, d, e, understood in terms of the successive orders of functions as nested radicals, explained in appendix B.]*

We can also assume that it is impossible to express $R^{\frac{1}{m}}$ by a rational function of the quantities $a$, $b$, etc. $p$, $p_1$, $p_2$, and by putting $\frac{R}{p_1^m}$ in place of $R$, it is clear that we can make $p_1 = 1$. Then

$$y = p + R^{\frac{1}{m}} + p_2 R^{\frac{2}{m}} + \cdots + p_{m-1} R^{\frac{m-1}{m}}. \qquad [A3]$$

*[Abel simplifies [A2] by getting rid of $p_1$: he redefines $R \to \frac{R}{p_1^m}$ so that $p_1 R^{\frac{1}{m}} \to p_1 \left( \frac{R}{p_1^m} \right)^{\frac{1}{m}} = R^{\frac{1}{m}}$. In what follows, Abel assumes that this has been done to remove $p_1$.]*

Substituting this value for $y$ in the equation proposed [A1] and reducing, we obtain a result of this form:

$$P = q + q_1 R^{\frac{1}{m}} + q_2 R^{\frac{2}{m}} + \cdots + q_{m-1} R^{\frac{m-1}{m}}, \qquad [A4]$$

$q$, $q_1$, $q_2$, etc. being rational and entire functions [i.e., polynomials] of the quantities $a$, $b$, $c$, $d$, $e$, $p$, $p_2$, ... and $R$.

*[Now Abel substitutes the presumed solution [A3] back into the main equation, [A1]. Doing so leads to [A4], where the new functions $q, q_1, q_2, \ldots$ depend on all the previous quantities: the coefficients $a, b, c, d, e$ and the quantities he has just used, $p, p_1, p_2, \ldots, R$. Note that since [A1] sets a polynomial to zero, this rewritten equation also does so, $P = 0$. Note also that, for convenience, Abel has defined $q, q_1, q_2, \ldots$ so that each multiplies the appropriate power of $R^{\frac{1}{m}}$. Note also that the highest power of $R$ in [A4] will be $\frac{m-1}{m}$, as shown in appendix B, pp. 173–174.]*

In order for this equation to be valid, it is necessary that $q = 0, q_1 = 0, q_2 = 0, \ldots, q_{m-1} = 0$.

In fact, calling $R^{\frac{1}{m}} = z$, we have two equations

$$z^m - R = 0 \quad \text{and} \quad q + q_1 z + \cdots + q_{m-1} z^{m-1} = 0. \quad [A5]$$

*[Abel sets out to prove $q = 0, q_1 = 0, q_2 = 0, \ldots, q_{m-1} = 0$; his proof requires several steps, ending with [A8]. He begins by defining $z = R^{\frac{1}{m}}$, so that $z^m = R$ or $z^m - R = 0$. Also, in [A4] substitute $z = R^{\frac{1}{m}}$, giving $q + q_1 z + \cdots + q^{m-1} z^{m-1} = 0$. These two parts of [A5] constrain z.]*

If now the quantities $q, q_1, \ldots$ are not equal to zero, these equations [A5] necessarily have one or more common roots. If $k$ is the number of these [common] roots, we know that we can find an equation of degree $k$ that has as roots the $k$ roots mentioned and in which all the coefficients are rational functions of $R, q, q_1,$ and $q_{m-1}$. Let

$$r + r_1 z + r_2 z^2 + \cdots + r_k z^k = 0 \quad [A6]$$

be that equation. It has these roots in common with the equation $z^m - R = 0$; thus, all the roots of this equation have the form $\alpha_\mu z$, where $\alpha_\mu$ designates one of the roots of the equation $\alpha_\mu^m - 1 = 0$. Then substituting [into [A6] $z \to \alpha_\mu z$], we have the following equations,

$$r + r_1 z + r_2 z^2 + \cdots + r_k z^k = 0 \qquad \text{[A7a]}$$

$$r + \alpha r_1 z + \alpha^2 r_2 z^2 + \cdots + \alpha^k r_k z^k = 0 \qquad \text{[A7b]}$$

. . . . . .

$$r + \alpha_{k-2} r_1 z + \alpha_{k-2}^2 r_2 z^2 + \cdots + \alpha_{k-2}^k r_k z^k = 0. \qquad \text{[A7k]}$$

*[In $z = \alpha_\mu z$, the subscript $\mu$ is an index that goes from 1 to k (the total number of common roots in [A5], $z$, $\alpha z$, $\alpha_1 z$, $\alpha_2 z$, ..., $\alpha_{k-2} z$). Substituting $z = \alpha_\mu z$ back into $z^m - R = 0$ yields $\alpha_\mu^m z^m - R = 0$, but $z^m = R$ and so this means that $\alpha_\mu^m R - R = 0$. Dividing by R (which $\neq 0$) shows that $\alpha_\mu$ is a root of the equation $\alpha_\mu^m - 1 = 0$. Now $\alpha_\mu$ stands for the whole series of values $1, \alpha, \alpha_2, ..., \alpha_{k-2}$, where Abel has noted that 1 is a root of this equation and calls $\alpha_1 = \alpha$. (Note that there are then only $k - 2$ values of $\alpha_\mu$ since, of the k roots, the first and second are 1 and $\alpha$.) Abel now substitutes the successive values $\alpha_\mu z = z, \alpha z, \alpha_2 z, ..., \alpha_{k-2} z$ back into [A6]. Substituting 1 for $\alpha_\mu$ gives [A7a]; substituting $\alpha$ gives [A7b], ... until substituting $\alpha_{k-2}$ yields [A7k].]*

From these $k$ equations, one can always find the value of $z$ expressed by a rational function of the quantities $r, r_1, r_2, ..., r_k$, and, as these quantities are themselves rational functions of $a, b, c, d, e, R, ..., p, p_2, ...$, it follows that $z$ is also a rational function of these same quantities, which is contrary to the hypothesis. Therefore, it is necessary that

$$q = 0, q_1 = 0, ..., q_{m-1} = 0. \qquad \text{[A8]}$$

*[Now Abel makes the critical observation: from the k equations [A7a–A7k], we can always find $z$ as a rational function of $r, r_1, ..., r_k$ and $\alpha$ because we have k simultaneous equations to determine the k unknowns, $z_1, z_2, ..., z_k$. Notice that this is different*

*from the situation with the original equation [A1], which was one equation to determine five values of y; [A7a–A7k] are k linear equations to determine k unknowns, and can be solved by elimination: Treat each power of z as a separate unknown and solve the k equations as if they were simultaneous linear equations for those k unknowns. But we assumed that $z = R^{\frac{1}{m}}$ is not a rational function of its variables, so we are forced to the only alternative, $q = q_1 = q_2 = \cdots = q_{m-1} = 0$, concluding the proof of [A8].]*

If now these equations are valid, it is clear that the proposed equation [A1] is satisfied by all the values that one obtains for y by giving to $R^{\frac{1}{m}}$ all the values

$$R^{\frac{1}{m}}, \alpha R^{\frac{1}{m}}, \alpha^2 R^{\frac{1}{m}}, \alpha^3 R^{\frac{1}{m}}, \ldots, \alpha^{m-1} R^{\frac{1}{m}}, \qquad [A9]$$

$\alpha$ being a root of the equation

$$\alpha^{m-1} + \alpha^{m-2} + \cdots + \alpha + 1 = 0. \qquad [A10]$$

*[If $y_1 = p + R^{\frac{1}{m}} + p_2 R^{\frac{2}{m}} + \cdots + p_{m-1} R^{\frac{m}{m-1}}$ [A3], then $q + q_1 R^{\frac{1}{m}} + \cdots + q_{m-1} R^{\frac{m-1}{m}} = 0$. This happens because, in substituting this form $y_1$ into [A1], we get terms like (products of p, $p_2, \ldots$) $(R^{\frac{1}{m}})^a (R^{\frac{2}{m}})^b \cdots$. Collecting powers, this becomes (products of p, $p_2, \cdots$) $(R^{\frac{1}{m}})^{a+2b+\cdots}$. Since the exponent $a + 2b + \cdots$ can always be written as $mi + j$, where i, j are integers ($j \leq m - 1$), then the integral powers of $R^i = (R^{\frac{1}{m}})^{mi}$ can be factored out and included with the products of p, $p_2, \cdots$ as $q, q_1, \cdots$ in [A4]. Abel has just shown that all these qs are zero. Now if we consider $y_2$, in which $R^{\frac{1}{m}} \to \alpha R^{\frac{1}{m}}$, a similar argument applies, except that there is now a factor of $\alpha$ raised to some power $(a + 2b + \cdots)$ multiplied times these previous factors of q and R. But since the qs are zero, then each term still vanishes and this $y_2$ also satisfies $P = 0$ in [A4]. The same argument also applies for $y_3$ ($R^{\frac{1}{m}} \to \alpha^2 R^{\frac{1}{m}}$) and all the other values of [A9].]*

We see also that all these values of $y$ are different, for in the contrary case we would have an equation of the same form as the equation $P = 0$, and such an equation leads, as we have seen, to a result that cannot be valid. The number $m$ thus cannot exceed 5. Therefore designating by $y_1, y_2, y_3, y_4$, and $y_5$ the roots of the proposed equation [A1], we will have

$$y_1 = p + R^{\frac{1}{m}} + p_2 R^{\frac{2}{m}} + \cdots + p_{m-1} R^{\frac{m-1}{m}}, \qquad \text{[A11a]}$$

$$y_2 = p + \alpha R^{\frac{1}{m}} + \alpha^2 p_2 R^{\frac{2}{m}} + \cdots + \alpha^{m-1} p_{m-1} R^{\frac{m-1}{m}}, \qquad \text{[A11b]}$$

$$\cdots \cdots$$

$$y_m = p + \alpha^{m-1} R^{\frac{1}{m}} + \alpha^{m-2} p_2 R^{\frac{2}{m}} + \cdots + \alpha p_{m-1} R^{\frac{m-1}{m}}. \qquad \text{[A11m]}$$

*[Note that, if on the contrary $y_1 = y_2$ (for example), then [A11a] = [A11b], which would require that $\alpha - 1 = 0 = \alpha^2 - 1 = \alpha^3 - 1 = \cdots$, which contradicts [A10]. Therefore all the values of $y$ are different. Of course, there are special cases of quintics having equal roots, but they will prove to have solutions. For instance, if all the roots are equal, $y = y_0$, then the quintic equation can just be factored into $(y - y_0)^5 = 0$, and clearly that can only hold if the coefficients are highly restricted. Abel also reminds us that the roots can be no more than five in number.]*

From these equations, we easily deduce that

$$p = \frac{1}{m}(y_1 + y_2 + \cdots + y_m), \qquad \text{[A12a]}$$

$$R^{\frac{1}{m}} = \frac{1}{m}(y_1 + \alpha^{m-1} y_2 + \cdots + \alpha y_m), \qquad \text{[A12b]}$$

$$p_2 R^{\frac{2}{m}} = \frac{1}{m}(y_1 + \alpha^{m-2} y_2 + \cdots + \alpha^2 y_m), \qquad \text{[A12c]}$$

$$\cdots \cdots$$

$$p_{m-1} R^{\frac{m-1}{m}} = \frac{1}{m}(y_1 + \alpha y_2 + \cdots + \alpha^{m-1} y_m). \qquad \text{[A12m]}$$

We see from this that $p, p_2, \ldots, p_{m-1}, R$, and $R^{\frac{1}{m}}$ are rational functions of the [roots of the] proposed equation [A1].

*[Now he calls these five roots $y_1, y_2, \ldots, y_5$ and uses [A3] and the result of [A9–A10] to write them out explicitly in [A11a–A11m]. Then he adds these equations up and gets*

$$y_1 + y_2 + \cdots + y_m = mp + (1 + \alpha + \alpha^2 + \cdots + \alpha^{m-1})R^{\frac{1}{m}}$$
$$+ p_2(1 + \alpha + \alpha^2 + \cdots + \alpha^{m-1})R^{\frac{2}{m}} + \cdots$$
$$+ p_{m-1}(1 + \alpha + \alpha^2 + \cdots + \alpha^{m-1})R^{\frac{m-1}{m}}. \qquad [A12.1]$$

*But from [A10] this leads immediately to [A12a], since all the sums $(1+\alpha+\alpha^2+\cdots+\alpha^{m-1})$ vanish. Now Abel will tease out the other terms in [A4] by multiplying each equation in such a way as to isolate each term, as follows. To find out what $R^{\frac{1}{m}}$ is, multiply [A11a] by 1, [A11b] by $\alpha^{m-1}$, [A11c] by $\alpha^{m-2}$, \ldots, [A11m] by $\alpha$. Then add them up. We get:*

$$y_1 + \alpha^{m-1}y_2 + \alpha^{m-2}y_3 + \cdots + \alpha y_m = (1 + \alpha + \alpha^2 + \cdots + \alpha^{m-1})p$$
$$+ m\alpha^m R^{\frac{1}{m}} + p_2\alpha^m(1 + \alpha + \alpha^2 + \cdots + \alpha^{m-1})R^{\frac{2}{m}} + \cdots$$
$$+ \alpha^m p_{m-1}(1 + \alpha + \alpha^2 + \cdots + \alpha^{m-1})R^{\frac{m-1}{m}}. \qquad [A12.2]$$

*Since $\alpha^m = 1$, then $mR^{\frac{1}{m}} = y_1 + \alpha^{m-1}y_2 + \alpha^{m-2}y_3 + \cdots + \alpha y_m$ [A12b]. Exactly similar tricks yield the rest of [A12b–m].]*

Let us now consider one of these quantities, for example $R$. Let

$$R = S + v^{\frac{1}{n}} + S_2 v^{\frac{2}{n}} + \cdots + S_{n-1}v^{\frac{n-1}{n}}. \qquad [A13]$$

Treating this quantity in the same manner as $y$, we obtain a similar result showing that the quantities $v^{\frac{1}{n}}, v, S, S_2, \ldots$ are rational functions of the different values of the function $R$, and, as these [values] are rational functions of $y_1, y_2$, etc., the functions $v^{\frac{1}{n}}, v, S, S_2, \ldots$ are also.

Pursuing this reasoning, we conclude that all the irrational functions contained in the expression for $y$ are rational functions of the roots of the proposed equation.

*[This concludes Step II of the whole proof.]*

This being established, it is not hard to complete the demonstration. Let us first consider the irrational functions of the form $R^{\frac{1}{m}}$, $R$ being a rational function of $a, b, c, d$, and $e$. Let $R^{\frac{1}{m}} = r$, [where] $r$ is a rational function of [the roots] $y_1, y_2, y_3, y_4$, and $y_5$ and $R$ is a symmetric function of these quantities. Now since the case in question is the general solution of the fifth-degree equation, it is clear that one can consider $y_1, y_2, y_3, y_4$, and $y_5$ as independent variables; thus the equation $R^{\frac{1}{m}} = r$ can hold under this supposition. Consequently, we can interchange the quantities $y_1, y_2, y_3, y_4$, and $y_5$ among themselves in the equation $R^{\frac{1}{m}} = r$, since by this interchange $R^{\frac{1}{m}}$ necessarily takes $m$ different values since $R$ is a symmetric function.

*[Now the final stage of the* reductio *begins. Abel lets $R^{\frac{1}{m}} = r$ and reminds us that he has just shown it to be a rational function of the roots $y_1, y_2, \ldots, y_5$. Furthermore, $R$ is a symmetric function of these roots. This means that we can permute the roots among themselves without changing $R$. It also means that the equation $R^{\frac{1}{m}} = r$ can be permuted among the m roots, and r will then assume the m different values, however they are permuted.]*

The function $r$ must also take $m$ different values in permuting in all ways possible the five variables it contains. To show this, it is necessary that $m = 5$ or $m = 2$, since $m$ is a prime number. (See the memoir by M. Cauchy in the *Journal de l'école polytechnique*, vol. 17.)

*[See appendix C for Abel's expansion of this argument. Cauchy's theorem states that if $m = 5$, our function $r = R^{\frac{1}{m}}$ can take on only five or two values, never three or four values. Cauchy's theorem*

*allows r to take only one value, but that would contradict our initial assumption that all the roots are different.]*

First, let $m = 5$. The function $r$ therefore has five different values and can consequently be put in the form

$$R^{\frac{1}{5}} = r = p + p_1 y_1 + p_2 y_1^2 + p_3 y_1^3 + p_4 y_1^4, \qquad [A14]$$

$p, p_1, p_2, \ldots$ being symmetric functions of $y_1, y_2, \ldots$. This equation gives, on interchanging $y_1$ and $y_2$,

$$p + p_1 y_1 + p_2 y_1^2 + p_3 y_1^3 + p_4 y_1^4$$
$$= \alpha p + \alpha p_1 y_2 + \alpha p_2 y_2^2 + \alpha p_3 y_2^3 + \alpha p_4 y_2^4 \qquad [A15]$$

where

$$\alpha^4 + \alpha^3 + \alpha^2 + \alpha + 1 = 0. \qquad [A16]$$

But this equation is impossible, so $m$ consequently must equal 2.

*[In his 1826 paper, Abel gives a much fuller account of these terse assertions, as well as a much simpler argument for the impossibility of $m = 5$. First, the simple argument: Consider expressing one of the roots $y_1$ as in [A11a], $y_1 = p + R^{\frac{1}{5}} + p_2 R^{\frac{2}{5}} + \cdots + p_4 R^{\frac{4}{5}}$ and then derive the expression for $R^{\frac{1}{5}} = \frac{1}{5}(y_1 + \alpha^4 y_2 + \alpha^3 y_3 + \alpha^2 y_4 + \alpha y_5)$, as in [A12b], where the case $m = 5$ has been taken. However, this relation is impossible, since the left-hand side has five values (the possible values of the fifth root) while the right-hand side has 120 (the permutations of the five roots). Therefore the case $m = 5$ is excluded.*

*Alternatively, here is Abel's longer reasoning for [A14–A15]: Consider a rational function $v$ that depends on $y_1, y_2, y_3, y_4, y_5$ and is symmetric under permutations of four out of five of these, $y_2, y_3, y_4, y_5$. Then we can express $v$ in terms of $y_1$ and of the coefficients of an equation of which $y_2, y_3, y_4, y_5$ are the solutions. To* *see this, write* $(y - y_2)(y - y_3)(y - y_4)(y - y_5) = y^4 - py^3 + qy^2 - ry + s$. *The roots* $y_2$, $y_3$, $y_4$, $y_5$ *can be expressed in terms of the coefficients* $p, q, r, s$, *since this is a quartic equation and hence solvable. Now* $y_1$, $y_2$, $y_3$, $y_4$, $y_5$ *are all the roots of a quintic equation* $y^5 - ay^4 + by^3 - cy^2 + dy - e = 0$ *[A1], which means that* $(y - y_1)(y - y_2)(y - y_3)(y - y_4)(y - y_5) = y^5 - ay^4 + by^3 - cy^2 + dy - e = (y - y_1)(y^4 - py^3 + qy^2 - ry + s)$, *as was just assumed. Then we can compare these forms of the same equation and conclude that* $p, q, r, s$ *can be expressed in terms of the root* $y_1$ *and the original coefficients* $a, b, c, d$. *(In detail, writing out this last expression and factoring gives* $(y - y_1)(y^4 - py^3 + qy^2 - ry + s) = y^5 - (p + y_1)y^4 + (q + py_1)y^3 - (r + qy_1)y^2 + (s + ry_1)y - sy_1$, *so that* $p = a - y_1, q = b - ay_1 + y_1^2, r = c - by_1 + ay_1^2 - y_1^3, s = d - cy_1 + by_1^2 - ay_1^3 + y_1^4$.)*

*Thus, the function* $v$ *can be expressed rationally in terms of* $y_1, a, b, c, d$. *Then it follows that* $v$ *can be written as a series of terms* $v = r_0 + p_1 y_1 + p_2 y_1^2 + \cdots + p_m y_1^m$, *where* $p_0, p_1, \ldots, p_m$ *are polynomial functions of* $a, b, c, d, e$, *the dependence on* $y_1$ *having been factored out (here Abel uses an argument very similar to that in appendix B to argue that the denominator can always be incorporated into the rational terms* $p_0, p_1, \ldots, p_m$*). Since* $y_1$ *is a root of [A1], then* $y_1^5 = ay_1^4 - by_1^3 + cy_1^2 - dy_1 + e$, *which can be used to re-express all the terms in* $v = p_0 + p_1 y_1 + p_2 y_1^2 + \cdots + p_m y_1^m$ *in which m is 5 or greater.*

*Thus, we can write any function that is symmetric under permutations of* $y_2$, $y_3$, $y_4$, $y_5$ *in the form* $v = p_0 + p_1 y_1 + p_2 y_1^2 + p_3 y_1^3 + p_4 y_1^4$.

*Now if in any such function we permute all five of* $y_1$, $y_2$, $y_3$, $y_4$, $y_5$, *then it follows that it takes only one value and is thus symmetric (since it is already symmetric in* $y_2$, $y_3$, $y_4$, $y_5$ *and now is also in* $y_1$*) or it takes five values (one for each of* $y_1 \rightarrow y_1$, $y_2$, $y_3$, $y_4$, $y_5$. *Thus (since we exclude the symmetric case), this argument shows that any function that has five different values* $y_1$, $y_2$, $y_3$, $y_4$, $y_5$ *must* *take the form* $v = p_0 + p_1 y_1 + p_2 y_1^2 + p_3 y_1^3 + p_4 y_1^4$ *[A14], and similarly for the other possible four values* $y_2$, $y_3$, $y_4$, $y_5$. *(Abel discusses various cases in which* $v$ *takes different numbers of values, but this is the essence of his argument.)*

*We also know that these different roots are related by [A9],* $R^{\frac{1}{5}} \rightarrow R^{\frac{1}{5}}$, $\alpha R^{\frac{1}{5}}$, $\alpha^2 R^{\frac{1}{5}}$, $\alpha^3 R^{\frac{1}{5}}$, $\alpha^4 R^{\frac{1}{5}}$. *Then take [A14] in the form* $R^{\frac{1}{5}} = r = p + p_1 y_2 + p_2 y_2^2 + p_3 y_2^3 + p_4 y_2^4$ *and multiply both sides by* $\alpha$, *so as to take* $R^{\frac{1}{5}} \rightarrow \alpha R^{\frac{1}{5}}$ *and hence* $y_1 \rightarrow y_2$. *By doing this, we get* $\alpha(p + p_1 y_1 + p_2 y_1^2 + p_3 y_1^3 + p_4 y_1^4) = p + p_1 y_2 + p_2 y_2^2 + p_3 y_2^3 + p_4 y_2^4$ *or (reversing the arbitrary names of* $y_1$ *and* $y_2$, *as Abel does)* $p + p_1 y_1 + p_2 y_1^2 + p_3 y_1^3 + p_4 y_1^4 = \alpha p + \alpha p_1 y_2 + \alpha p_2 y_2^2 + \alpha p_3 y_2^3 + \alpha p_4 y_2^4$, *which is [A15]. However, this equation cannot be satisfied unless either* $\alpha = 1$ *or* $y_1 = y_2$, *neither of which is allowed since the roots are all different. Therefore there can be only two values of* $r$, *by Cauchy's theorem.]*

Then let

$$R^{\frac{1}{2}} = r, \qquad\qquad\qquad\qquad\qquad\qquad [\text{A17}]$$

and then $r$ must have two different values of opposite sign. We then have (see the memoir of M. Cauchy)

$$R^{\frac{1}{2}} = r = v(y_1 - y_2)(y_1 - y_3) \cdots (y_2 - y_3) \cdots (y_4 - y_5)$$

$$= v S^{\frac{1}{2}}, \qquad\qquad\qquad\qquad\qquad\qquad [\text{A18}]$$

$v$ being a symmetric function.

*[Abel writes the two possible values for* $r$ *as [A17], remembering that the square root is always ambiguous up to a* $\pm$ *sign. Cauchy's theorem also implies in this case that* $r$ *can be written in the form [A18], where* $v$ *is a symmetric function (dependent on the coefficients) and* $S^{\frac{1}{2}}$ *is a special function,* $S^{\frac{1}{2}} = (y_1 - y_2)(y_1 - y_3) \cdots (y_4 - y_5)$ *discussed in appendix C. Note that* $S$ *cannot be zero because no two roots are the same.]*

Let us now consider irrational functions of the form

$$\left( p + p_1 R^{\frac{1}{\nu}} + p_2 R_1^{\frac{1}{\mu}} + \cdots \right)^{\frac{1}{m}},$$  [A19]

$p$, $p_1$, $p_2$, etc. $R$, $R_1$, etc. being rational functions of $a$, $b$, $c$, $d$, and $e$ and consequently symmetric functions of $y_1$, $y_2$, $y_3$, $y_4$, and $y_5$. As we have seen, we must have $\nu = \mu =$ etc. $= 2$, $R = v^2 S$, $R_1 = v_1^2 S$, etc. The preceding function [A19] can thus be written in the form

$$\left( p + p_1 S^{\frac{1}{2}} \right)^{\frac{1}{m}}.$$  [A20]

Let

$$r = \left( p + p_1 S^{\frac{1}{2}} \right)^{\frac{1}{m}},$$  [A21]

$$r_1 = \left( p - p_1 S^{\frac{1}{2}} \right)^{\frac{1}{m}}.$$  [A22]

Multiplying, we have

$$rr_1 = \left( p^2 - p_1^2 S \right)^{\frac{1}{m}}.$$  [A23]

*[Abel reminds us that we now know that we are dealing with $R^{\frac{1}{2}}$, rather than $R^{\frac{1}{5}}$. Using the definitions $R = v^2 S$, $R_1 = v_1^2 S$, ... he can express irrational functions such as [A19] in the form $r = (p + p_1 S^{\frac{1}{2}})^{\frac{1}{m}}$ [A21], where he has dropped out the other terms in [A19] since his object is merely to show us the general form they will take. Likewise, he writes another irrational term $r_1$ in the similar general form $r_1 = (p - p_1 S^{\frac{1}{2}})^{\frac{1}{m}}$ [A22]. Multiplying out $rr_1$ gives [A23]; he chooses the minus sign in defining $r_1$ to give this convenient form.]*

Now if $rr_1$ is not a symmetric function, $m$ must equal 2, but in this case $r$ would have four different values, which is impossible; therefore $rr_1$ must be a symmetric function.

*[Abel argues that the product $rr_1$ is a symmetric function; if it were not, then (by Cauchy's theorem) $m = 2$. What would this*

*mean? In [A21], r would have four values, since it would involve
a square root of terms including square roots. This is not allowed;
only m = 5 or m = 2 is possible. So then rr₁ is a symmetric
function, [A23].]*

Let $v$ be this [symmetric] function [$v = rr_1$], then

$$r + r_1 = \left(p + p_1 S^{\frac{1}{2}}\right)^{\frac{1}{m}} + v\left(p + p_1 S^{\frac{1}{2}}\right)^{-\frac{1}{m}} = z. \qquad [A24]$$

*[Remember that $r_1 = \frac{v}{r} = v\left(p + p_1 S^{\frac{1}{2}}\right)^{-\frac{1}{m}}$.]*

This function having $m$ different values, then $m$ must equal 5,
since $m$ is a prime number. Thus we have

$$z = q + q_1 y + q_2 y^2 + q_3 y^3 + q_4 y^4$$
$$= \left(p + p_1 S^{\frac{1}{2}}\right)^{\frac{1}{5}} + v\left(p + p_1 S^{\frac{1}{2}}\right)^{-\frac{1}{5}}, \qquad [A25]$$

$q, q_1, q_2$, etc. being symmetric functions of $y_1, y_2, y_3$, etc. and
thus rational functions of $a, b, c, d$, and $e$.

*[Since he has excluded the possibility that m = 2, by elimination,
m = 5. To regain contact with his quest for the roots of the original
equation, he writes z in [A25] in terms of the roots of the quintic,
$z = q + q_1 y + q_2 y^2 + q_3 y^3 + q_4 y^4$, using [A14].]*

Combining this equation with the proposed equation, we
can express $y$ in terms of a rational function of $z, a, b, c, d$,
and $e$. Now such a function is always reducible to the form

$$y = P + R^{\frac{1}{5}} + P_2 R^{\frac{2}{5}} + P_3 R^{\frac{3}{5}} + P_4 R^{\frac{4}{5}}, \qquad [A26]$$

where $P, R, P_2, P_3$, and $P_4$ are functions of the form $p + p_1 S^{\frac{1}{2}}$,
$p, p_1$, and $S$ being rational functions of $a, b, c, d$, and $e$.

*[Now he turns the expression [A25] "inside out," as he has done
many times before, to express y as a function of z and the coefficients
a, b, c, d, e. Because of [A11a], he can write this in the form of
[A26], where all the functions P, R, . . . can be written in the form
$p + p_1 S^{\frac{1}{2}}$ (because of the reasoning around [A20]).]*

From this expression for $y$ we derive that

$$R^{\frac{1}{5}} = \frac{1}{5}(y_1 + \alpha^4 y_2 + \alpha^3 y_3 + \alpha^2 y_4 + \alpha y_5)$$

$$= (p + p_1 S^{\frac{1}{2}})^{\frac{1}{5}}, \qquad\qquad\qquad\qquad [A27]$$

where

$$\alpha^4 + \alpha^3 + \alpha^2 + \alpha + 1 = 0. \qquad\qquad\qquad [A28]$$

*[For the last time, he turns [A26] inside out, to express $R^{\frac{1}{5}}$ in terms of the roots $y_1$, $y_2$, ..., yielding [A27]; this is just [A12b] again, with $m = 5$. He notes, crucially, that he has also established that $R^{\frac{1}{5}}$ can be equated to $(p + p_1 S^{\frac{1}{2}})^{\frac{1}{5}}$.]*

Now the left-hand side [A27] has 120 different values and the right-hand side has only 10; consequently, $y$ cannot have the form we have found, but we have proved that $y$ must necessarily have this form, if the proposed equation is solvable.

*[The left-hand side of [A27] has 120 different values because $y_1$ can take any of the five values of the five roots, leaving four possibilities for $y_2$, three possibilities for $y_3$, two possibilities for $y_4$, and only one for $y_5$. That means the total number of possible values of the left-hand side is $5! = 120$. But the right-hand side of [A27] has only ten possible values, since the square root has two possibilities, multiplied by the five possibilities for the fifth root. There is no way that something with 120 possible values can always equal something with only 10. Therefore, the premise fails that the equation can be solved algebraically, that is, by $y$ given by [A2].]*

Thus we conclude that *it is impossible to solve in radicals the general equation of the fifth degree.*

It follows immediately from this theorem that it is also impossible to solve in radicals general equations of degrees higher than the fifth.

*[Quite simply, if you multiply an unsolvable quintic by y, it becomes a sixth-degree equation, with one root y = 0 and the other roots unsolvable, and likewise for higher degrees. In terms of Abel's proof, the same argument that he has given will continue to apply when n > 5 since Cauchy's theorem still holds. For instance, consider an equation of degree 7. Then Cauchy's theorem says that any function in the solution can take only 7, 2, or 1 values, and we can reapply all of Abel's arguments that these will all lead to contradictions exactly parallel to those for degree 5. It is worth considering why this doesn't affect degrees 2, 3, and 4. For them, the number of roots matches the possible values for functions in the solution (respectively 2, 3, 4). Only for degree 5 or higher does the gap open between the number of roots and the permitted number of values for the functions in the solution (5, 2, 1, but never 3 or 4.)]*

# Appendix B
# Abel on the General
# Form of an Algebraic
# Solution

Abel needs to prove that the general solution $y$ of an equation has the form

$$y = p + p_1 R^{\frac{1}{m}} + p_2 R^{\frac{2}{m}} + \cdots + p_{m-1} R^{\frac{m-1}{m}}, \qquad \text{[B1 = A2]}$$

where $m$ is a prime number and $R$, $p$, $p_1$, $p_2$, etc. are functions of the same form as $y$, perhaps containing further nested radicals inside them but ultimately (after a finite number of radicals within radicals) containing only rational functions of the coefficients of the original equation. The basic idea is simple: We assume that a solution has a finite number of terms, involving a finite number of radicals. What Abel proves is that, however complex such an expression might be, we can always arrange it in the form [B1]. The main idea is to make all the terms (whether inside radicals or coefficients) rational by always putting them over a common denominator, a messy but straightforward process. He did not spell out this argument in his 1824 paper, so we follow here the account in his 1826 paper.

First, Abel considers a polynomial function, $f(x)$, which is a sum of terms, each $x$ raised to some power. A rational function is defined as the quotient of two polynomial functions, similar to the way a rational number is the quotient of two integers. Abel means by "algebraic function" one that can be expressed by rational functions along with taking roots, a special case of the modern use of the term "algebraic function" for all polynomials $f(x, y) = 0$. He now sets up a hierarchy of complexity and defines an algebraic function of the "zeroth order" as $f(a, b, c, d, e)$, where $f$ is a rational function of the coefficients $a, b, c, d, e$. Then an algebraic function of the "first order" is $f(p_0^{\frac{1}{m}})$, where $p_0$ is a function of the zeroth order and $f$ is a rational function of the coefficients in zeroth order. Similarly, a "second order" function $f(p_1^{\frac{1}{m}})$ involves first-order functions $p_1$ and $f$. We can keep going to define an algebraic function of the $k$th order that has $k$ levels of roots-within-roots; we call this quite general algebraic function $v$. By factoring $m$ and writing $R^{\frac{1}{m}}$ as a sequence of successive radicals, we may assume that $m$ is prime.

We can now write the algebraic function $v$ as $\frac{T}{V}$, where $V$ is expressed in the form $V = v_0 + v_1 R^{\frac{1}{n}} + v_2 R^{\frac{2}{n}} + \cdots + v_m R^{\frac{m}{n}}$, and likewise for $T$. (Here we have reverted to Abel's notation in the 1824 paper.) Now Abel considers the series of substitutions that take $R^{\frac{1}{n}} \to \alpha R^{\frac{1}{n}}$, $R^{\frac{1}{n}} \to \alpha^2 R^{\frac{1}{n}}$, ..., $R^{\frac{1}{n}} \to \alpha^{n-1} R^{\frac{1}{n}}$, where $\alpha \neq 1$ but satisfies $\alpha^n = 1$, as in [A10]. Under these $n - 1$ substitutions, $V$ will in general take the $n - 1$ values $V_1, V_2, \ldots, V_{n-1}$. If we multiply the definition $v = \frac{T}{V}$ by $1 = \frac{V_1 V_2 \ldots V_{n-1}}{V_1 V_2 \ldots V_{n-1}}$, then

$$v = \frac{T V_1 V_2 \ldots V_{n-1}}{V V_1 V_2 \ldots V_{n-1}}. \qquad [B2]$$

Abel now shows that the denominator of [B2] is a polynomial. To see this, remember that $V = v_0 + v_1 R^{\frac{1}{n}} + v_2 R^{\frac{2}{n}} + \cdots + v_m R^{\frac{m}{n}}$. Under the substitution $R^{\frac{1}{n}} \to \alpha R^{\frac{1}{n}}$, $V \to V_1 = v_0 + \alpha v_1 R^{\frac{1}{n}} + \alpha^2 v_2 R^{\frac{2}{n}} + \cdots + \alpha^m v_m R^{\frac{m}{n}}$; when $R^{\frac{1}{n}} \to \alpha^2 R^{\frac{1}{n}}$, $V$

becomes $V_2 = v_0 + \alpha^2 v_1 R^{\frac{1}{n}} + \alpha^4 v_2 R^{\frac{2}{n}} + \cdots + \alpha^{2m} v_m R^{\frac{m}{n}}$; and so forth. To find the denominator of [B2], we multiply these out and then gather similar terms: $V V_1 V_2 \cdots V_{n-1} = (v_0 + v_1 R^{\frac{1}{n}} + v_2 R^{\frac{2}{n}} + \cdots + v_m R^{\frac{m}{n}}) \times (v_0 + \alpha v_1 R^{\frac{1}{n}} + \alpha^2 v_2 R^{\frac{2}{n}} + \cdots + \alpha^m v_m R^{\frac{m}{n}}) \times (v_0 + \alpha^2 v_1 R^{\frac{1}{n}} + \alpha^4 v_2 R^{\frac{2}{n}} + \cdots + \alpha^{2m} v_m R^{\frac{m}{n}}) \times \cdots \times (v_0 + \alpha^{n-1} v_1 R^{\frac{1}{n}} + \alpha^{2(n-1)} v_2 R^{\frac{2}{n}} + \cdots + \alpha^{m(n-1)} v_m R^{\frac{m}{n}}) = v_0^n + \alpha v_0^{n-1} v_1 (1 + \alpha + \alpha^2 + \cdots + \alpha^{n-1}) + \alpha^2 v_0^{n-2} v_2 R^{\frac{2}{n}} (1 + \alpha + \alpha^2 + \cdots + \alpha^{n-1}) + \cdots = v_0^n$, where we have collected the factors of each power of $R^{\frac{1}{n}}$, $R^{\frac{2}{n}}$, ..., $R^{\frac{m}{n}}$, and noted that $(1 + \alpha + \alpha^2 + \cdots + \alpha^{n-1}) = 0$ [A10], so that only $v_0^n$ remains. If $v_0$ is a polynomial, we are done. If $v_0$ contains further radicals nested inside it, this same process can be repeated as many times as required until we finally reach polynomials, for we know that there is only a finite number of subradicals. Thus, the denominator of [B2] is a polynomial and all the irrational functions are in the numerator. Furthermore, the rational functions in the numerator can be redefined to include the polynomial denominator.

This is the crucial realization. By the same reasoning, the numerator is a polynomial in terms of $R^{\frac{1}{n}}$ and powers of the variables, so that it can be written as

$$v = q_0 + q_1 R^{\frac{1}{n}} + q_2 R^{\frac{2}{n}} + \cdots + q_m R^{\frac{m}{n}}, \qquad [B3]$$

a function of order $k$, where $R, q_0, q_1, \ldots$ are functions that are of order $k - 1$ and include the factors absorbing the polynomial denominator. In the case of the quintic equation, $R, q_0, q_1, \ldots$ would all be functions of order 4, that is, roots of a quartic equation. In [B3], we can always assume that $m < n$, for otherwise we can write $m = an + b$, where $a$ and $b < n$ are integers. Then

$$R^{\frac{m}{n}} = R^{\frac{an+b}{n}} = R^a R^{\frac{b}{n}}. \qquad [B4]$$

We can absorb the integral powers of the polynomial $R^a$ into a redefined coefficient $q'_b = q_b R^a$. This leaves the fractional powers of $R$ expressed in terms of lower-order fractions, $R^{\frac{b}{n}}$. Because we have factored $k$ with $a, b < n$, the highest power of $R^{\frac{1}{n}}$ is less than $n$, namely $R^{\frac{n-1}{n}}$. This shows that [B4] will take the stated form [B1], as claimed.

Let us consider some examples. Box 6.1 shows that the quadratic equation obeys this form. The case of the cubic equation $y^3 + a_1 y - a_0 = 0$ is only slightly more complicated. The Cardano solution of this equation (see box 2.4) is

$$y = \sqrt[3]{\frac{a_0}{2} + \sqrt{\frac{a_0^2}{4} + \frac{a_1^3}{27}}} + \sqrt[3]{\frac{a_0}{2} - \sqrt{\frac{a_0^2}{4} + \frac{a_1^3}{27}}}. \qquad [B5]$$

To show that this really is of Abel's form, write $y = p + R^{\frac{1}{3}} + p_2 R^{\frac{2}{3}}$. Since our cubic equation has no $y^2$ term, the sum of the roots is 0; but the sum of the roots is also $3p + R^{\frac{1}{3}}(1 + \alpha + \alpha^2) + R^{\frac{2}{3}}(1 + \alpha + \alpha^2) = 3p$, since $1 + \alpha + \alpha^2 = 0$. Hence $p = 0$.

Now substitute $y = R^{\frac{1}{3}} + p_2 R^{\frac{2}{3}}$ into the cubic equation and factor out the various powers of $R$. One gets:

$$(R + p_2^3 R^2 - a_0) + R^{\frac{1}{3}}(3p_2 R + a_1) + R^{\frac{2}{3}}(3R p_2^2 + a_1 p_2) = 0. \qquad [B6]$$

By [A8] in appendix A, this means that the separate co-efficients are each zero: $R + p_2^3 R^2 - a_0 = 0$, $3p_2 R + a_1 = 0$, and $3R p_2^2 + a_1 p_2 = 0$. Thus $p_2 = -\frac{a_1}{3R}$, and by substitution, $R^2 - a_0 R - \frac{a_1^3}{27} = 0$. One solution of this quadratic is $R_+ = \frac{a_0}{2} + \sqrt{\frac{a_0^2}{4} + \frac{a_1^3}{27}}$. Then $y = R_+^{\frac{1}{3}} + p_2 R_+^{\frac{2}{3}} = R_+^{\frac{1}{3}} - \frac{a_1}{3} R_+^{-\frac{1}{3}}$. But $R_+^{\frac{1}{3}} R_-^{\frac{1}{3}} = -\frac{a_1}{3}$, where $R_- = \frac{a_0}{2} - \sqrt{\frac{a_0^2}{4} + \frac{a_1^3}{27}}$. Thus $y = R_+^{\frac{1}{3}} + R_-^{\frac{1}{3}}$, which is exactly the form shown in [B5]. Therefore the solution of the cubic equation can be written in Abel's form.

# Appendix C
# Cauchy's Theorem on Permutations

Abel's 1824 paper uses Cauchy's argument about permutations, which in turn generalizes Ruffini's results. In his 1826 paper, Abel gives a further account, summarized here and related to the modern language of group theory. Cauchy considers any function of $n$ variables, $f(x_1, x_2, x_3, x_4, x_5)$ (here for the case $n = 5$). If the value of the function is unchanged when the values of the variables are permuted, it is said to be a *symmetric* function. That is, if the function is symmetric, it does not matter which variable is which: $f(x_1, x_2, x_3, x_4, x_5) = f(x_2, x_1, x_3, x_4, x_5) = f(x_5, x_2, x_3, x_4, x_1) = \cdots$, and so on through all the $5! = 120$ possible permutations of the five variables. If the function is not symmetric, different permutations of $x_1, x_2, x_3, x_4, x_5$ will lead to different values of the function. In general, the greatest number of such different values of the function cannot exceed the total number of permutations, $n!$. Call the values $f(A_1), f(A_2), \ldots, f(A_{n!})$, where $A_1, A_2, \ldots, A_{n!}$ are all the permutations of the original order $x_1, x_2, x_3, \ldots$.

Assume that the functions takes $p$ different values, where $p < n!$. First we show that $p$ must be a divisor of $n!$. Then some of the $n!$ values must be equal. Let us suppose that the first $m$ values are equal, $f(A_1) = f(A_2) = \cdots = f(A_m)$. In this series, let us make the successive substitutions $A_1 \to A_{m+1}, A_2 \to A_{m+2}, \ldots, A_m \to A_{2m}$. This gives another $m$ equal values (though in general not equal to the first $m$). We can then continue this process of substitution, finally exhausting the $n!$ possibilities, dividing them into $p$ sets, each having $m$ elements. Therefore, $pm = n!$ and so $p$ is a divisor of $n!$.

In modern group theory, this is called Lagrange's Theorem. Consider a group $G$ and call the number of its elements $|G|$ (where $|G| = n!$ if $G$ is the group of permutations of $n$ elements). Consider any subgroup $H$ contained in $G$. Lagrange's Theorem states that the number of elements of $H$, namely $|H|$, must be a divisor of $|G|$. Let $H$ be the subgroup of permutations that fix one of the $p$ possible values of our function, say $f(A_1)$. Then the procedure described above corresponds to partitioning the group $G$ into "cosets of $H$ in $G$." (If the group $G$ is noncommutative, "left" cosets are distinguished from "right.") The argument just given shows that these cosets partition the elements of $G$ into exhaustive, exclusive classes of $p$ elements and so $p$ must be a divisor of $|G| = n!$. Since the cosets divide up $G$ into these exhaustive and exclusive classes, it is natural to think of them as the "quotient $G/H$," which is a group when $H$ is a normal subgroup (defined on pp. 193–194).

A good example of this is the arithmetic of a clock. On a 24-hour clock, each numeral is a coset, representing an equivalence class of each hour: 13 o'clock is equivalent to $13 + 24 = 37$ o'clock, $13 + 48 = 61$ o'clock, etc. If we then shift to a 12-hour clock, we take the "quotient" of the 24-hour clock divided by 2, since now 13 and 1 o'clock are also equivalent, as are 25 and 37 o'clock, and so forth. This is only a brief clarification; for more examples and details, see the works listed in the notes, pp. 191–192.

In general, finite groups can be compared to molecules that can be analyzed by breaking them into smaller molecules and ultimately into their constituent atoms. As the same chemical atoms can form different molecules, the same mathematical "atoms" can form different groups. Consider a group $G$ that contains a normal subgroup $H$ that is "proper" ($H$ is neither simply $G$ nor the identity). Then the quotient group $G/H$ and $H$ are constituent "sub-molecules" of $G$. Continue breaking these "sub-molecules" down in the same way into ever smaller pieces, until we cannot break them down further: the collection of indecomposable pieces that we get are the "atoms" of the group $G$. Jordan and Hölder's famous theorem proves that this collection of "atoms" is determined entirely by the group $G$. From this standpoint, Abel's proof means that an equation is solvable in radicals if and only if the "atoms" of its group are all abelian (in particular, cyclic and having orders that are prime numbers, if the group is finite).

We began by considering a symmetric function of $n$ variables, which has only one value when we permute the variables. Other functions may have two values. For example, consider the function that Cauchy and Abel call

$$\sqrt{S} = (x_1 - x_2)(x_1 - x_3)(x_1 - x_4)(x_1 - x_5)(x_2 - x_3) \cdots (x_4 - x_5),$$

in which all five variables appear *antisymmetrically*, meaning that if we make the permutation $x_1 \rightarrow x_2$, the function changes sign but is equal in absolute magnitude: $\sqrt{S} \rightarrow -\sqrt{S}$. In fact, any interchange of two variables will produce two possible values for the function, opposite in sign. We can also, of course, multiply this antisymmetric function by a symmetric function, which will not change upon permutation (Abel uses just this function in his [A18], appendix A).

It is *not* always possible to make a function that has any given number of values larger than 2, even if it is a divisor of $n!$. In what follows, let $p$ be the largest prime less than or equal to $n$. Let $A_m$ be any permutation of order $p$, meaning that when it is applied $p$ times we return to the starting point, which we will write as $f(A_m)^p = f(A_m)^0$. If we assume that the function takes on *fewer* than $p$ values, then two among those values must be equal, say $f(A_m)^r = f(A_m)^{r'}$, where $0 < r, r' < p - 1$. Then apply the permutation $A_m$ to both sides of this equation $(p - r)$ times: $f(A_m)^{r+p-r} = f(A_m)^{r'+p-r}$, or $f(A_m) = f(A_m)^j$, since $(r + p - r) = p$, $f(A_m)^p = f(A_m)$, and we define $j = r' + p - r$. Not only are two of the values equal, $f(A_m) = f(A_m)^j$, but also $f(A_m) = f(A_m)^{bj}$, where $b$ is any integer, each multiple of $j$ representing that many times around the circle. Since $p$ is a prime number, we can always find two integers $a, b$ such that $bj = pa + 1$. Hence $f(A_m)^{bj} = f(A_m)^{pa+1} = f(A_m)$, which we showed is also equal to $f(A_m) = f(A_m)^{pa}$. This means that adjacent values of the function are equal, $f(A_m)^{pa} = f(A_m)^{pa+1}$, so that the permutation $A_m$ does not change the value of the function. Therefore, Cauchy concludes, if the function takes fewer values than $p$, it is not changed by a permutation of order $p$.

Let us imagine a function that takes fewer values than $p$ (assuming now that $p > 3$, as will be true in the case of the quintic, for which $n = p = 5$). From our last result, we know that this function is unchanged by a permutation of order $p > 3$. Now to make such a permutation, we must transpose variables. For example, to go from $f(x_1, x_2, x_3, x_4, x_5)$ to $f(x_2, x_1, x_3, x_4, x_5)$, we must transpose $x_1 \leftrightarrow x_2$. Likewise, any arbitrary permutation is composed of a certain number of transpositions between the variables. This number can be even or odd.

Now consider two permutations of order $p$: In the first, let $x_a \to x_b$, $x_b \to x_c$, $x_c \to x_d$, ..., $x_p \to x_a$, replacing each variable with the next in the list. To use the notation introduced in chapter 8, this is $(abcd \cdots p)$. Clearly, this is a substitution of order $p$, for after $p$ such transpositions, we return to the starting point. Similarly, consider a second series: $x_b \to x_c$, $x_c \to x_a$, $x_d \to x_b$, $x_e \to x_d$, $x_f \to x_e$, ..., $x_a \to x_p$ or $(ap \cdots edbc)$. This too is a substitution of order $p$, for after $p$ such substitutions, we would come around to the starting position. Indeed, it is a slightly varied version of the first substitution (moving $x_a$ to the end). So both will leave the function unchanged, by the result in the preceding paragraph. Now notice what happens when we perform the first and then the second substitution in that order: $x_a \to x_b \to x_c$, $x_b \to x_c \to x_a$, $x_c \to x_d \to x_b$, $x_e \to x_f \to x_e$, $x_f \to x_g \to x_f$, ..., $x_p \to x_a \to x_p$ or $(abcd \cdots p) * (ap \cdots edbc) = (acb)$. After the first three terms, the variables remain unchanged after both substitutions are performed. So these two permutations performed in sequence yield $x_a \to x_c$, $x_b \to x_a$, $x_c \to x_b$, with all the other variables unchanged. Though these examples may seem arbitrary, they are not. Any permutation of order $p$ must take the form of these examples, for we could have used any combination of variables in place of $x_a$, $x_b$, $x_c$, .... The upshot is that, in general, given any 3-cycle, we can always find two permutations of order $p$ ($p$-cycles, as they are called in chapter 8) whose product is the given 3-cycle.

Since neither permutation of order $p$ can change the value of the function (by our earlier result), neither can both of them, and thus neither can a permutation of order 3. We now show that a permutation of order 3 can always be written as the result of two permutations of order 2, which are just transpositions. Consider the one we discussed earlier, $x_a \to x_c$, $x_b \to x_a$, $x_c \to x_b$. It can be written as the transposition

$x_a \leftrightarrow x_b$ followed by the transposition $x_b \leftrightarrow x_c$. To check this, do them in the order specified, $x_a \rightarrow x_b \rightarrow x_c$, $x_b \rightarrow x_a$, $x_c \rightarrow x_b$, giving us the same result: $(ab) * (bc) = (acb)$. Again, this is quite general in form, so any permutation of order 3 is equivalent to two permutations of order 2 (transpositions). But we just learned that the function is unchanged by a permutation of order 3; therefore, it is unchanged by the product of two transpositions. Similarly, it is unchanged by an even number of transpositions. To prove this, it suffices to show that any two distinct transpositions can be written as two 3-cycles. Consider, then, $x_a \leftrightarrow x_b$ followed by $x_c \leftrightarrow x_d$. It can also be accomplished by $x_a \rightarrow x_b \rightarrow x_c$ followed by $x_c \rightarrow x_a \rightarrow x_d$, verifying the assertion: $(ab) * (cd) = (abc) * (cad)$. Thus, since two transpositions (or any even number) do not change the function, each transposition changes the sign of the function. Two transpositions change the sign twice, restoring the original sign, whereas any odd number of transpositions must change the sign of the function. They are antisymmetric, to use the term introduced earlier.

In either case, the conclusion is that the function we have been discussing can have at most two values. Remember that all this followed *if $p \leq n$*. So finally we reach the main theorem of Cauchy that is so important for Abel: *The number of different values a function of n values can take cannot be lower than the largest prime number $p \leq n$ without becoming equal to 2.* Note that, if $n = 5$, $p = 5$ also, so that in the case of the quintic, either the function has five values or it has at most two values. In this form, Cauchy and Ruffini's insight is crucial to the last stage of Abel's proof.

# Notes

## 1  The Scandal of the Irrational

Pythagoras of Samos: Scholars have questioned every aspect of the traditional account. See Walter Burkert, *Lore and Science in Ancient Pythagoreanism*, tr. Edwin L. Minar Jr. (Cambridge, MA: Harvard University Press, 1972) and also C. A. Huffman, "The Pythagorean Tradition," in *The Cambridge Companion to Early Greek Philosophy*, ed. A. A. Long (Cambridge: Cambridge University Press, 1999), 66–87. Other accounts include Peter Gorman, *Pythagoras—A Life* (London: Routledge and Kegan Paul, 1979) and Leslie Ralph, *Pythagoras—A Short Account of His Life and Philosophy* (New York: Krikos, 1961). For a collection of original sources, see *The Pythagorean Sourcebook and Library*, ed. Ken Sylvan Guthrie (Grand Rapids, MI: Phares Press, 1987).

Euclid: The standard annotated edition is *The Thirteen Books of Euclid's Elements*, tr. Thomas L. Heath (New York: Dover, 1956), which gives the proof of incommensurability cited by Aristotle (*Prior Analytics* 41a, 26–27) at 3:2. For an engaging modern encounter with this classic, see Robin Hartshorne, *Geometry: Euclid and Beyond* (New York: Springer-Verlag, 2000).

Pythagorean mathematics and the golden ratio: See H. E. Huntley, *The Divine Proportion: A Study in Mathematical Beauty* (New York: Dover, 1970).

*Ṛta*: A. L. Basham, *The Wonder That Was India* (New York: Grove Press, 1954), 113, 236–237. For the Indo-European roots, see Julius Pokorny, *Indogermanisches Etymologisches Wörterbuch* (Bern: Francke Verlag, 1959), 1:56–60. I thank Eva Brann for drawing my attention to the connection between *arithmos* and "rite."

Plato: The episode of the slave boy is in *Meno* 84d–85b. See also Jacob Klein, *A Commentary on Plato's Meno* (Chapel Hill, NC: University of North Carolina Press, 1965), 103–107, especially his point that, although the propositions came from Socrates, "the *assent* and the *rejection* came from nobody but the boy himself" (105). For Meno's character, see Xenophon, *Anabasis* 2: 21–29. The "young irrationals" are discussed in Plato's *Republic*, Book VII, 534d. For the story of the black and white horses, see *Phaedrus* 253d–254e; for the story of Theaetetus, see *Theaetetus* 142a–150b; Socrates as midwife, 150c–151d. A *poros* is a means of passage, like a bridge or ferry, so *aporia* signifies being stuck, unable to cross over. For a helpful discussion of Theodorus and Theaetetus, see Wilbur Richard Knorr, *The Evolution of the Euclidean Elements* (Dordrecht: D. Reidel, 1975), 62–108, which includes a plausible account why Theodorus stopped at 17 (181–193) and a critique of earlier explanations (109–130).

Greek mathematics: For a helpful survey, see the classic book by Carl B. Boyer, *A History of Mathematics*, second ed., revised by Uta C. Merzbach (New York: John Wiley, 1991), 43–99 (especially 72–74 on incommensurability), 100–119 (Euclid), cited hereafter as HM. Another essential classic is Jacob Klein, *Greek Mathematical Thought and the Origin of Algebra*, tr. Eva Brann (New York: Dover, 1992), 3–113. See also Morris Kline, *Mathematical Thought from Ancient to Modern Times* (New York: Oxford University Press, 1972), 28–34, hereafter cited as MT.

Pappus on the irrational: *The Commentary of Pappus on Book X of Euclid's Elements*, tr. William Thomson (Cambridge, MA: Harvard University Press, 1930), 64–65, translation revised thanks to Bruce Perry.

Pentagons and irrationality: See Kurt von Fritz, "The Discovery of Incommensurability by Hippasus of Metapontum," *Annals of Mathematics* 46, 242–264 (1945).

Touchstone and torture: I have discussed the issue of experiment as the "torture" of nature in my book *Labyrinth: A Search for the Hidden Meaning of Science* (Cambridge, MA: MIT Press, 2000), chapter 2, in which I argue that Francis Bacon did not say or mean that nature was to be abused by the experimental trials of the new science. Though Bacon did not use the charged term "torture" in this context, Plato uses it to describe the arduous ordeal of dialectic and inquiry; however, from the context it is clear that he understands it to be noble, not base or abusive. This also applies to his daring metaphor of philosophical inquiry as "parricide"; see my paper "Desire, Science, and Polity: Francis Bacon's Account of Eros," *Interpretation* 26:3, 333–352 (1999), note 10. For the later thinkers, see my "Wrestling with Proteus: Francis Bacon and the 'Torture' of Nature," *Isis* 90:1, 81–94 (1999),

"Nature on the Rack: Leibniz' Attitude towards Judicial Torture and the 'Torture' of Nature," *Studia Leibnitiana* 29, 189–197 (1998), and "Proteus Unbound: Francis Bacon's Successors and the Defense of Experiment," *Studies in Philology* 98:4, 428–456 (2001).

Euclid on irrationals: See *The Elements*, Book X, propositions 1 (on the indefinite divisibility of any magnitude) and 115 (the infinite number of kinds of irrationals). For a general survey of the context of numbers and irrationality, see Midhat Gazalé, *Number: From Ahmes to Cantor* (Princeton: Princeton University Press, 2000). For the development of the Greek theory of ratio, see Sir Thomas Heath, *A History of Greek Mathematics* (New York: Dover, 1981), 1:90–91, 154–157, and Howard Stein, "Eudoxos and Dedekind: On the Ancient Greek Theory of Ratios and Its Relation to Modern Mathematics," *Synthese* 84, 163–211 (1990).

Greek music: See M. L. West, *Ancient Greek Music* (Oxford: Clarendon Press, 1992), 233–242, and the invaluable collection of original texts with commentary in *Greek Musical Writings*, ed. Andrew Barker (Cambridge: Cambridge University Press, 1989), 1:137, 188, 411, 419, including also Euclid's musical treatise in 2:190–208. I have found no passage where an ancient writer noted the near equality of $\sqrt{2}$ and the tritone, which is understandable given their primal assumption that musical interval is inherently a ratio.

## 2   Controversy and Coefficients

Babylonian mathematics: HM, 23–42; MT, 3–14. For an interesting comparison of Greek and Babylonian approaches to mathematics, see Simone Weil, *Seventy Letters*, tr. Richard Rees (London: Oxford University Press, 1965), 112–127.

Chinese algebra: For a general overview, see Joseph Needham, *Science and Civilisation in China* (Cambridge: Cambridge University Press, 1959), 3:1–170, with a comment on the cubic problem on 125–126. For a discussion of an important text, with valuable excerpts and translations, see J. Hoe, "The Jade Mirror of the Four Unknowns—Some Reflections," *Mathematical Chronicle* 7, 125–156 (1978), who argues that the "inherent symbolism of the Chinese language" moves in the direction of modern algebraic symbolism.

Al-Khwārizmī: See my *Labyrinth*, chapter 7. His work (called in Arabic *Al-jabr wa'l muqabālah*) is included in *A Source Book in Mathematics, 1200–1800*, ed. D. J. Struik (Cambridge, MA: Harvard University Press, 1969), 55–60. See also B. L. van der Waerden, *A History of Algebra from al-Khwārizmī to Emmy Noether* (New York: Springer-Verlag, 1980). For an excellent survey, see Karen Hunger Parshall, "The Art of Algebra from al-Khwārizmī to Viète: A Study in the Natural Selection of Ideas," *History of Science* 26, 129–164 (1988). There is a valuable collection of essays in Roshdi Rashed, *The Development of Arabic Mathematics: Between Arithmetic and Algebra* (Dordrecht: Kluwer Academic, 1994).

Khayyām: For a valuable collection of translations with helpful commentary, see R. Rashed and B. Vahabzadeh, *Omar Khayyam the Mathematician* (New York: Bibliotheca Persica, 2000). Rashed argues that Omar the poet and Omar the mathematician may not be the same person; his assertion remains controversial.

Shylock on commerce: *The Merchant of Venice*, I.iii.18–21.

Fibonacci: See HM, 254–257, John Fauvel and Jeremy Gray, *The History of Mathematics: A Reader* (London: Macmillan, 1987), 241–243, and J. Gies and F. Gies, *Leonard of Pisa and the New Mathematics of the Middle Ages* (New York: Crowell, 1969).

$e$: See Eli Maor, $e$: *The Story of a Number* (Princeton: Princeton University Press, 1998). If \$1 is invested in a bank account at 100% per annum, compounded annually, at the end of one year its value will be \$2. But if the interest is compounded at every instant, continuously, at the end of one year the account will be worth $e = 2.718\dots$ dollars.

Pacioli: R. Emmett Taylor, *No Royal Road: Luca Pacioli* (New York: Arno Press, 1980), contains a translation of the portions of the *Summa* relating to double-entry bookkeeping, as does B. S. Yamey, *Luca Pacioli's Exposition of Double-Entry Bookkeeping; Venice 1494* (Venice: Abrizzi, 1994), 9–33, 95–171. See also R. G. Brown and K. S. Johnston, *Pacioli on Accounting* (New York: McGraw-Hill, 1965), and Richard H. Macve, "Pacioli's Lecacy," in *Accounting History from the Renaissance to the Present: A Remembrance of Luca Pacioli*, ed. T. A. Lee, A. Bishop, R. H. Parker (New York: Garland, 1996), 3–30. For the connections with the development of printing, see E. L. Eisenstein, *The Printing Press as an Agent of Social Change: Communications and Cultural Transformations in Early-Modern Europe* (Cambridge: Cambridge University Press, 1979), 548. Pacioli includes idealized fonts based on geometry at the end of his *De divina proportione* (Venice: 1507; reprint Maslianico: Dominioni, 1967).

Leonardo da Vinci and Pacioli: See Carlo Zammattio, Augusto Marinoni, and Anna Maria Brizio, *Leonardo the Scientist* (New York: McGraw-Hill, 1980), 88–117, and Emanuel Winternitz, *Leonardo da Vinci as a Musician* (New Haven: Yale University Press, 1982), 10–16.

Piero della Francesca: For an outstanding account of his great mathematical achievements, see Mark A. Peterson, "The Geometry of Piero della Francesca," *Mathematical Intelligencer*, 19:3, 33–40 (1997), which also clarifies Pacioli's plagiarisms from Piero's works and discusses the relation of algebra and geometry in Piero's time.

History of double-entry bookkeeping: See Michael Chatfield, *A History of Accounting Thought* (Huntington, NY: Krieger Publishing, 1977), 3–18, who emphasizes that "the significance of the … integration of real and nominal accounts [in double-entry bookkeeping] far surpasses every other aspect of accounting development." For the connection between economic and mathematical developments, see also Frank Swetz, *Capitalism and Arithmetic: The New Math of the 15th Century* (La Salle, IL: Open Court, 1987).

Cardano: See Girolamo Cardano, *Ars magna, or the Rules of Algebra*, tr. T. Richard Witmer (New York: Dover, 1993), which unfortunately renders Cardano's mathematical expressions into modern algebraic notation, instead of retaining his own rhetorical mode of expression that does not rely on such symbolism. For his colorful autobiography, see Girolamo Cardano, *The Book of My Life* (New York: Dover, 1962), and also Øystein Ore, *Cardano, the Gambling Scholar* (Princeton: Princeton University Press, 1953), which treats the cubic and quartic equations on 59–107 (the commercial problem cited is on 69–70), giving very helpful excerpts from the invective of the combatants and also the complete text of Cardano's *Book on Games of Chance* (first printed posthumously in 1663). See also Anthony Grafton, *Cardano's Cosmos: The Worlds and Works of a Renaissance Astrologer* (Cambridge, MA: Harvard University Press, 2000).

Tartaglia's poem: Fauvel and Gray, *History of Mathematics*, 255–256, which also includes other documents from the controversy on 253–263.

Solutions of cubic and quartic equations: For a complete and lucid account, see the classic account by J. V. Uspensky, *Theory of Equations* (New York: McGraw-Hill, 1948), 82–98. William Dunham gives an engaging presentation in *Journey through Genius: The Great Theorems of Mathematics* (New York: John Wiley, 1990), 133–154. See also HM, 282–288, and MT, 263–272. My account of completing the cube draws on Parshall, "The Art of Algebra," 144–147. For a neat exposition of the factorization of these equations, see Morton J. Hellman, "A Unifying Technique for the Solution of the Quadratic, Cubic, and Quartic," *American Mathematical Monthly* 65, 274–276 (1958); Hellman shows the failure of this technique to the quintic in "The Insolvability of the Quintic Re-Examined," *American Mathematical Monthly* 66, 410 (1959).

"Nature does not permit it": Cardano, *Ars magna*, 9.

Viète and the "new algebra": The seminal work is Klein, *Greek Mathematical Thought*, 161–185; I cite Viète's "Introduction to the Analytical Art" from the translation in this volume by J. Winfree Smith (315–353), which discusses John Wallis's legal account of "species" at 321–322, n. 10. See also Helena M. Pycior, *Symbols, Impossible Numbers, and Geometric Entanglements* (Cambridge: Cambridge University Press, 1997), 27–39. I have discussed the connections between Viète's work in codebreaking and algebra in my *Labyrinth*, 73–83, and (in greater detail) in "Secrets, Symbols, and Systems: Parallels between Cryptanalysis and Algebra, 1580–1700," *Isis* 88, 674–692 (1997). For translations of the original Viète documents, see my paper "François Viète, Father of Modern Cryptanalysis—Two New Manuscripts," *Cryptologia* 21:1, 1–29 (1997).

History of symbolic notation: See Florian Cajori, *A History of Mathematical Notations* (Mineola, NY: Dover, 1993), 117–123 (Cardano), 181–187 (Viète).

# 3   Impossibilities and Imaginaries

Van Roomen's problem: See Viète's response, "Ad Problema, qvod omnibvs mathematicis totivs orbis contruendum proposuit Adrianus Romanus" (1595), in his *Opera Mathematica* (Hildesheim: Georg Olms, 2001), 305–324, discussed in HM, 309–311, and also by Guido Vetter, "Sur l'équation du quarante-cinquième degré d'Adriaan van Roomen," *Bulletin des sciences mathématiques* (2) 54, 277–283 (1954). Viète had derived expressions for the sines or cosines of multiple angles, such as $\sin n\theta = n\cos^{n-1}\theta\sin\theta - \frac{n(n-1)(n-2)}{3\cdot2\cdot1}\cos^{n-3}\theta\sin^3\theta + \cdots$, where the subsequent signs alternate and the coefficients are the alternate numbers in the arithmetic triangle (Pascal's triangle). Viète set $K = \sin45\theta$, which his formula could then express in terms of $x = 2\sin\theta$, yielding van Roomen's equation (see box 2.7). To solve it, he found $\theta = (\arcsin K)/45$, giving $x = 2\sin\theta$.

Kepler: See my paper "Kepler's Critique of Algebra," *Mathematical Intelligencer* 22:4, 54–59 (2000), which gives the full quotations and sources, and also my *Labyrinth*, 91–94. For Kepler's treatment of the heptagon, see Johannes Kepler, *The Harmony of the World*, tr. E. J. Aiton, A. M. Duncan, and J. V. Field (Philadelphia: American Philosophical Society, 1997), 60–79, and D. P. Walker, "Kepler's Celestial Music," in his *Studies in Musical Science in the Late Renaissance* (Leiden: E. J. Brill, 1978), 34–62; J. V. Field, *Kepler's Geometrical Cosmology* (Chicago: University of Chicago Press, 1988), 99–105; J. V. Field, "The Relation between Geometry and Algebra: Cardano and Kepler on the Regular Heptagon," in *Girolamo Cardano: Philosoph, Naturforscher, Arzt*, ed. E. Kessler (Wiesbaden: Harrassowitz Verlag, 1994), 219–242. For further discussion of the heptagon and its relation to angle trisection, see Andrew Gleason, "Angle Trisection, the Heptagon, and the Triskaidecagon," *American Mathematical Monthly*, 95, 185–194 (1988).

Galileo on the Book of Nature: See "The Assayer" in *Discoveries and Opinions of Galileo*, tr. Stillman Drake (New York: Doubleday, 1957), 237–238. For Galileo's attitude toward mathematics, see Carl B. Boyer, "Galileo's Place in the History of Mathematics," in *Galileo, Man of Science*, ed. Ernan McMullin (New York: Basic Books, 1967), 232–255.

Negative and imaginary quantities: HM, 219–220 (negative numbers in Hindu mathematics), 276–278 (Chuquet), 287–289 (Bombelli), 305–306 (Girard). See also Pycior, *Symbols, Impossible Numbers*. For Gauss's account of negatives and imaginaries, see the excellent anthology *From Kant to Hilbert: A Source Book in the Foundations of Mathematics*, ed. William B. Ewald (Oxford: Clarendon Press, 1996), 1:310–313, 307 (Leibniz on the amphibian). For an engaging account of the developing conception of imaginary numbers, see Barry Mazur, *Imagining Numbers (particularly the square root of minus fifteen)* (New York: Farrar Straus Giroux, 2002). See also Paul J. Nahin, *An Imaginary Tale: The Story of $\sqrt{-1}$* (Princeton: Princeton University Press, 1998).

Girard: See his 1629 treatise *Invention nouvelle en l'algèbre*, in *The Early Theory of Equations* (Annapolis, MD: Golden Hind Press, 1986).

Bombelli: For a thoughtful introduction that includes translated selections, see Federica La Nave and Barry Mazur, "Reading Bombelli," *Mathematical Intelligencer* 24:1, 12–21 (2002).

Descartes: See *The Geometry of René Descartes*, tr. David Eugene Smith and Marcia L. Latham (New York: Dover, 1954), 174–175 (introduction of "imaginary" roots), 160–161 (Descartes's rule of signs), 162–175 (methods for increasing or scaling the value of roots; Descartes's "relativity"), 22–37 (the locus problem for three or more lines), 176–192 (methods for solving higher-degree equations), 220–239 (solution of a special sixth-degree equation, using conic sections), 240–241 ("the general method to construct all problems"; the pleasure of discovery). For a discussion of Descartes's concept of number, see Klein, *Greek Mathematical Thought*, 197–211.

## 4   Spirals and Seashores

Newton: For his *Lectures on Algebra, 1673–1683*, see *The Mathematical Papers of Isaac Newton*, ed. D. T. Whiteside (Cambridge: Cambridge University Press, 1972), 5:130–135, 565 (two sample passages showing his treatment of equations). Quotations in the text are from Isaac Newton, *The* Principia: *Mathematical Principles of Natural Philosophy*, tr. I. Bernard Cohen and Anne Whitman (Berkeley: University of California Press, 1999), 485 (geometrical synthesis), 483–485 (Newton's solution to locus problem), 511–513 (lemma 28). For further discussion of the significance and validity of lemma 28, see my paper "The Validity of Newton's Lemma 28," *Historia Mathematica* 28, 215–219 (2001), and also Bruce Pourciau, "The Integrability of Ovals: Newton's Lemma 28 and Its Counterexamples," *Archive for the History of Exact Sciences*, 55, 479–499 (2001). For the relation of Newton's mathematical preferences to his other projects, see my *Labyrinth*, 113–133.

Tschirnhaus, Bring, and Jerrard: HM, 432–434; for Jerrard's mistaken solution, see chapter 9 notes, below.

Leibniz on the quintic equation: See the helpful historical overview in Nicolas Bourbaki (the pseudonym of a group of French mathematicians), *Elements of the History of Mathematics*, tr. John Meldrum (New York: Springer-Verlag, 1994), 69–80 at 74.

Fundamental Theorem of Algebra: For a complete treatment of all of Gauss's proofs, see Benjamin Fine and Gerhard Rosenberger, *The Fundamental Theorem of Algebra* (New York: Springer-Verlag, 1997), especially 182–186 on Gauss's original proof. Their treatment is based on Uspensky, *Theory of Equations*, 293–297, which I also follow in my exposition. See also Heinrich Dörrie, *100 Great Problems of Elementary Mathematics*, tr. David Antin (New York: Dover, 1965), 108–112. There is a nice one-page proof of this theorem by Uwe F. Mayer, "A Proof That Polynomials Have Roots," *College Mathematics Journal* 28:1, 58 (1999).

# 5 Premonitions and Permutations

Lagrange and Vandermonde: See the classic sturdy by Hans Wussing, *The Genesis of the Abstract Group Concept*, tr. Abe Shenitzer (Cambridge, MA: MIT Press, 1984), 71–79, van der Waerden, *History of Algebra*, 76–83, and MT, 600–606, which suggests at 605 that Lagrange was "drawn to the conclusion that the solution of the general higher-degree equation (for $n > 4$) by algebraic equations was likely to be impossible," though Lagrange never expressed this opinion explicitly. His paper "Réflexions sur la résolution algébrique des équations" (1770–1771) is in *Œuvres de Lagrange* (Paris: Gauthier-Villars, 1869), 3:205–422. Citations in the text are taken from van der Waerden, *History of Algebra*, 81.

G. E. Montucla (1725–1799):  Cited from the excellent article by Raymond G. Ayoub, "Paolo Ruffini's Contributions to the Quintic," *Archive for History of Exact Sciences* 23, 253–277 (1980), at 257.

Gauss:  W. K. Bühler, *Gauss: A Biographical Study* (New York: Springer-Verlag, 1981), 36 (Gauss's lack of interest in combinatorics).

Gauss on the quintic: Carl Friedrich Gauss, *Disquisitiones Arithmeticae*, tr. Arthur A. Clarke (New Haven: Yale University Press, 1965), 445 (a passage added in proof, according to Bühler, *Gauss*, 77), 407–460 (general discussion of the problem of equal division of a circle, also known as the cyclotomic equation). For a helpful simplified discussion of the 17-gon and the cyclotomic equation, see Dörrie, *100 Great Problems*, 177–184. There is also an excellent account of this in Felix Klein, *Famous Problems of Elementary Geometry*, tr. W. W. Beman and David Eugene Smith (New York: Chelsea, 1962), 24–41.

Ruffini:  See Wussing, *Group Concept*, 80–84, van der Waerden, *History of Algebra*, 83–85, and particularly Ayoub, "Paolo Ruffini's Contributions to the Quintic," including the quote from Ruffini at 263 ("behold a very important theorem"). Another useful article is R. A. Bryce, "Ruffini and the Quintic Equation," in *First Australian Conference on the History of Mathematics*, ed. John N. Crossley (Clayton, Victoria: Department of Mathematics, Monash University, 1981), 5–31, which points out a number of errors in Ruffini's work that substantiate the attribution of the theorem to Abel alone. To my knowledge, none of Ruffini's papers has been translated into English; they are available in the original Italian in *Opere Matematiche di Paolo Ruffini* (Palermo: Tipografia Matematica di Palermo, 1915–1953), 2 volumes.

# 6   Abel's Proof

For a comprehensive account of Abel's life and its Norwegian context, see Arild Stubhaug, *Niels Henrik Abel and His Times* (Berlin: Springer-Verlag, 2000), 178–183 (the arrival of Holmboe), 239–240 (Abel's false solution to the quintic), 297 (Abel and Fermat's Last Theorem). Still useful is the briefer, earlier work by Øystein Ore, *Niels Henrik Abel: Mathematician Extraordinaire* (Minneapolis: University of Minnesota Press, 1957). There are interesting essays also in the volume marking the centenary of Abel's birth, *Niels Henrik Abel: Mémorial publié à l'occasion du centénaire de sa naissance* (Kristiania: Jacob Dybwad, 1902).

Fermat's Theorem:  For Gauss's proof in the case $n = 3$, see Dörrie, *100 Great Problems*, 96–104; for a popular history through Andrew Wiles's 1993 proof, see Simon Singh, *Fermat's Enigma* (New York: Walker, 1997). For Abel's work on this problem, see his 1826 letter to Holmboe, cited in *Œuvres Complètes de Niels Henrik Abel*, 2:254–255.

Abel's Theorem: Modern treatments tend to subsume Abel's own work in Galois theory. Some older textbooks give an account of Abel's work, though sometimes in an unfamiliar notation; see the citations below under Galois Theory and J. Pierpont, "On the Ruffini-Abelian Theorem," *Bulletin of the American Mathematical Society* 2, 200–221 (1896). For helpful accounts phrased in the language of modern algebra, see Lars Gårding and Christian Skau, "Niels Henrik Abel and Solvable Equations," *Archive for the History of Exact Sciences* 48, 81–103 (1994) and Michael I. Rosen, "Niels Hendrik Abel and Equations of the Fifth Degree," *American Mathematical Monthly* 102, 495–505 (1995). Dörrie, *100 Great Problems*, 116–127, gives an account of Leopold Kronecker's clarified version of the theorem.

Abel and Ruffini: The quote is my translation from Abel's 1828 paper "Sur la resolution algebrique des équations," *Œuvres Complètes de Niels Henrik Abel,* ed. L. Sylow and S. Lie (New York: Johnson Reprint, 1965), 2:217–243 at 218. Sylow notes (1:293) that Abel first commented on Ruffini in 1826 in an unsigned note in Férussac's *Bulletin*.

# 7   Abel and Galois

Abel's later life: Stubhaug, *Abel*, 329–331 (Crelle), 395–420 (Abel in Paris), 468 (appeal of the four French mathematicians to the Swedish king). Quotes from Abel: 471 (monstrous egotists), 398–402 (billiards and theater), 424 (poorer than a church mouse), 474 (correspondence with Legendre), 471 (quite alone), 475–493 (Abel's death and his despair), 409 (on Cauchy).

Abel on commutativity: My translation of the opening of his 1828 paper "Mémoire sur une classe particulière d'équations résolubles algébriquement," in *Œuvres Complètes*, 2:478–507 at 478. See also William Snow Burnside and Arthur William Panton, *The Theory of Equations* (London: Longmans, Green, 1928), 282–305.

French history during the time of Abel and Galois: For a superb overview, see the classic work by Albert Guérard, *France: A Modern History* (Ann Arbor: University of Michigan Press, 1959), 3 (Capetian dynasty), 278–296 (the period 1814–1848), 288 (Lafayette).

Galois's life: The most reliable recent study is Laura Toti Rigatelli, *Évariste Galois 1811–1832* (Boston: Birkhäuser Verlag, 1996), which includes a helpful bibliography. Among earlier brief biographical works, Eric Temple Bell, *Men of Mathmatics* (New York: Simon & Schuster, 1986), 362–377, is widely known, though it propagates many of the myths concerning Galois; for poetic aspects of Abel and Galois, see also George Steiner, *Grammars of Creation* (New Haven: Yale University Press, 2001), 207–212. There is a helpful corrective in Tony Rothman, "Genius and Biographers: The Fictionalization of Évariste Galois," *American Mathematical Monthly* 89:2, 84–102 (1982) and "The Short Life of Évariste Galois," *Scientific American* 246:4, 136–149 (1982), also included in his *Science à la Mode: Physical Fashions and Fictions* (Princeton: Princeton University Press, 1989). See also the helpful articles by René Taton, "Évariste Galois and His Contemporaries," *Bulletin of the London Mathematical Society* 15, 107–118 (1983), and Harold M. Edwards, "A Note on Galois Theory," *Archive for the History of Exact Science* 41, 163–169 (1990). For Galois's final letter, see David Eugene Smith, *A Source Book in Mathematics* (New York: Dover, 1959), 278–285.

Raspail: For his account of Galois in prison, see Toti Rigatelli, *Galois*, 98–100; for his acquaintance with Abel, see Stubhaug, *Abel*, 410–411, 416–417 (Galois's comment on Abel's death).

# 8   Seeing Symmetries

My treatment of the relation between symmetries and geometric figures is inspired by Felix Klein's treatment of the symmetries of regular solids as paradigms of the symmetries of equations, found in a brief form in his classic lectures to schoolteachers, *Elementary Mathematics from an Advanced Standpoint*, tr. E. R. Hedrick and C. A. Noble (New York: Dover, n.d.), 1:101–143, and very fully (though with notation that is quite hard to follow) in his *Lectures on the Icosahedron* (New York: Dover, 1956), 3–20. I am not aware of other treatments that use the analogy with dance or the visualizations I present in the text and in box 8.3. I was greatly helped by Hartshorne's treatment of the symmetry groups of polyhedra in his *Geometry: Euclid and Beyond*, 469–480.

Basic treatments of Galois theory: My presentation attempts only to give an intuitive version (though see appendix C and the notes below for a few more details). There are many treatments; several stand out as especially accessible. John E. Maxfield and Margaret W. Maxfield, *Abstract Algebra and Solution by Radicals* (New York: Dover, 1992) is particularly suitable for self-study. Saul Stahl, *Introductory Modern Algebra: A Historical Approach* (New York: John Wiley, 1997) is an admirably written textbook that includes valuable excerpts from the writings of al-Khwārizmī, Cardano, Abel, Galois, and Cayley (261–287); I would also recommend Jean-Pierre Tignol, *Galois' Theory of Algebraic Equations* (London: Longman Scientific Technical, 1980). Charles Robert Hadlock, *Field Theory and Its Classical Problems* (Mathematical Association of America, 1978), despite its austere title, is quite accessible, though somewhat more abstract in its approach than the Maxfields' book. There is also a nice overview in I. M. Yaglom, *Felix Klein and Sophus Lie: Evolution of the Idea of Symmetry in the Nineteenth Century*, tr. Sergei Sossinsky (Boston: Birkhäuser, 1988), 1–21. Lillian R. Lieber, *Galois and the Theory of Groups: A Bright Star in Mathesis* (Lancaster, PA: Science Press, 1932), long out of print, gives a lucid outline in sixty pages, expressed in a kind of verse with drawings, a companion piece to her similarly delightful treatments of non-Euclidean geometry and general relativity.

Other treatments of Galois theory: A full listing of the vast number of treatments is scarcely possible here, but I would like to mention some that I found helpful. D. E. Littlewood gives an interesting overview in *The Skeleton Key of Mathematics: A Simple Account of Complex Algebraic Theories* (New York: Harper, 1960), 65–76, as does Kline, *Mathematical Thought*, 752–771. Old textbooks sometimes present the theory less abstractly: See Leonard E. Dickson, *Modern Algebraic Theories* (Chicago: Sanborn, 1926), 135–250, Burnside and Panton, *The Theory of Equations*, 244–305, and Edgar Dehn, *Algebraic Equations: An Introduction to the Theories of Lagrange and Galois* (New York: Dover, 1960). There is a nice account of the theory in an old French edition of Galois's works by G. Verriest, *Évariste Galois et la Théorie des Équations Algébriques* (Paris: Gauthier-Villars, 1934), which is included in *Œuvres Mathématiques d'Évariste Galois*, ed. Émile Picard (Paris: Gauthier-Villars, 1897). A more recent French treatment is Claude Mutafian, *Équations Algébriques et Théorie de Galois* (Saint-Amand-Montrond: Librarie Vuibert, 1980). German-speaking readers may find helpful N. Tschebotaröw, *Grundzüge der Galois'schen Theorie*, tr. H. Schwerdtfeger (Groningen: P. Noordhoff, 1950), which is thorough and rich in examples. Toti Rigatelli, *Galois*, 115–138, gives a valuable outline of his work. For more modern treatments that are friendly but still rigorous, see M. M. Postnikov, *Fundamentals of Galois Theory*, tr. Leo F. Boron (Groningen: P. Noordhoff, 1962), Ian Stewart, *Galois Theory* (London: Chapman and Hall, 1973), and R. Bruce King, *Beyond the Quartic Equation* (Boston: Birkhäuser, 1996), which is oriented to the symmetry concerns of chemists. Many graduate and undergraduate textbooks include Galois theory; see especially Nathan Jacobson, *Basic Algebra I*, second edition (New York: W. H. Freeman, 1985), 210–270, I. N. Herstein, *Topics in Algebra*, second edition (New York: Wiley, 1975), 237–259, and Jerry Shurman, *Geometry of the Quintic* (New York: Wiley, 1997), which concentrates on geometric aspects. For some helpful articles, see Raymond G. Ayoub, "On the Nonsolvability of the General Polynomial," *American*

*Mathematical Monthly* 89, 397–401 (1982), and John Stillwell, "Galois Theory for Beginners," *American Mathematical Monthly* 101, 22–27 (1994), both of which assume considerable familiarity with modern algebra. For the gradual interpretation of Galois's ideas, see Wussing, *Group Concept*, 118–141, van der Waerden, *History of Algebra*, 103–116, B. Melvin Kiernan, "The Development of Galois Theory from Lagrange to Artin," *Archive for the History of Exact Sciences* 8, 40–154 (1971–1972), MT, 752–771, and Yoïchi Hirano, "Note sur les diffusions de la théorie de Galois: Première clarification des idées de Galois par Liouville," *Historia Scientiarum* 27, 27–41 (1984).

Making models of Platonic solids: See David Mitchell, *Mathematical Origami: Geometrical Shapes by Paper Folding* (Norfolk: Tarquin Publications, 1999).

Groups: For a helpful introduction that stresses the larger philosophical significance of group theory, see Curtis Wilson, "Groups, Rings, and Lattices," *St. John's Review* 35, 3–11 (1985). In particular, Wilson notes that "the most general or universal aim of intellectual work is the discovery of invariants" (6).

Regular solids and their groups: For a classic work that includes a sketch of the symmetry argument for the uniqueness of the five Platonic solids, see Hermann Weyl, *Symmetry* (Princeton: Princeton University Press, 1980), 149–156.

Visualization of groups: Two books offer many ways to grasp group symmetries visually: Israel Grossman and Wilhelm Magnus, *Groups and Their Graphs* (Washington, D.C.: Mathematical Association of America, 1964), and R. P. Burn, *Groups: A Path to Geometry* (Cambridge: Cambridge University Press, 1985). These visualizations go back to Klein, *Lectures on the Icosahedron*, also treated by W. Burnside, *Theory of Groups of Finite Order*, second edition (Cambridge: Cambridge University Press, 1911), 402–408.

Normal subgroups: A subgroup $H$ of a group $G$ is called a *normal* or *invariant subgroup* of $G$ if for every element $g$ in $G$ and $h$ in $H$, there is some $h'$ such that $h' * g = g * h$, where $h'$ is not necessarily the same as $h$. Note that if we multiply both sides of this equation on the right by $g^{-1}$, the inverse of the element $g$, we find then $h' * g * g^{-1} = h' = g * h * g^{-1}$ (since by definition $g * g^{-1} = I$). All possible $g * h * g^{-1}$ are called the "conjugates" or the "equivalence class" of $h$. A normal subgroup need not be abelian but is always "self-conjugate": if it contains an element $h$, it also contains all the conjugates of $h$. This is the precise meaning of "elements of the same kind" on pp. 119, 122, 123–124.

Here is a more intuitive way of thinking about normal subgroups. Picture the elements $h$ of the subgroup $H$ as forming an "orbit," consisting of its possible values. Then the combined operation $g * h * g^{-1}$ tries to "tip" $H$ out of its orbit by imposing an "external" action $g$, balanced by an equal and opposite "counteraction" $g^{-1}$. If $g * h * g^{-1}$ always remains within its original orbit, for all actions $g$, the orbit is undisturbed and the subgroup is "normal." Also, if $a$ is a rotation around a vertex $A$, then $b = g * a * g^{-1}$ means that $b$ is the same kind of rotation around the vertex $B = g(A)$. (See Hartshorne, *Geometry: Euclid and Beyond*, 474.) Specifically, $g^{-1}$ takes the vertex $B$ back to $A$, then $a$ performs the rotation, and $g$ takes the vertex $A$ back to $B$. Thus, normal subgroups (which obey this condition) signify all the rotations of the same kind around all allowable vertices.

Jordan and Hölder: See Camille Jordan, *Traité des Substitutions* (Paris: Gauthier-Villars, 1957), 286–291 (abelian equations), 370–397 (composition series); and Wussing, *Group Concept*, 135 ff.

Solvable chains: Galois theory begins by determining a given equation's group of permutations. In practice, this can involve complex calculations. Given such a "Galois group," we can determine in sequence a maximal normal subgroup (a normal subgroup of largest order less than that of the group), a maximal normal sub-subgroup, and so on until we reach a sub-sub-subgroup that includes only the identity element, which is automatically a (normal) subgroup of any group, by definition. This chain can be written symbolically as $G \triangleright G_1 \triangleright G_2 \triangleright \cdots \triangleright G_{n-1} \triangleright G_n \triangleright I$, where $G_i \triangleright G_{i+1}$ means: "$G_i$ contains the normal subgroup $G_{i+1}$," though $G_{i+1}$ may *not* be normal in the whole group $G$.

Then the whole group $G$ is called *solvable* if and only if the quotient group of each successive group in the chain, $G_i / G_{i+1}$, is abelian. That is, if $H$ is a normal subgroup of $G$, then the quotient group $G/H$ is defined to be the set of all "cosets of $H$ in $G$," namely all the sets $\{g_1 * h_1, g_2 * h_1, \ldots\}, \{g_1 * h_2, g_2 * h_2, \ldots\}, \ldots$, made from all the elements $g_1, g_2, \ldots$ of $G$ and $h_1, h_2, \ldots$ of $H$. See appendix C, p. 176, for the example of clock arithmetic illustrating cosets and quotient groups. If the quotient group $G_i / G_{i+1}$ is nonabelian, the chain is broken, and the equation is not solvable in radicals.

Note also that if a finite group $G$ has a prime number of elements, $p$, then it is abelian. Proof: Let $g$ be an element of $G$ other than the identity (since $p > 1$). Then consider $H = \{I, g, g^2, \ldots\}$, which is a subgroup of order $n$, where $n > 1$. By Lagrange's Theorem (see appendix C), $n$ divides $p$. But $p$ is prime; it has no divisors besides itself and 1. Then if $n \neq 1$, $n$ must equal $p$ and $H$ is of the same order as $G$, so that $H = G$. $G$ is abelian since it is "cyclic," meaning that it is composed of powers of one element, $g$, namely

$\{I, g, g^2, \ldots\}$. Since $g^m * g^n = g^n * g^m$, any cyclic group is abelian, and so too is $G$. Note that not every abelian group is cyclic; for instance, the four-group $V$ (table 8.3) is abelian but not cyclic.

Monster group: See Stahl, *Introductory Modern Algebra*, 247–248; for a technical overview of the modern study of simple groups, see Ron Solomon, "On Finite Simple Groups and Their Classification," *Notices of the American Mathatical Society*, 42:2, 231–239 (1995). For a popular account, see W. Wayt Gibbs, "Monstrous Moonshine Is True," *Scientific American*, 279:5, 40–41 (1998).

# 9   The Order of Things

History of noncommutativity: For an excellent selection of original writings with commentary, see Ewald, *From Kant to Hilbert*, 1:293–296 (Gauss on commutative law), 314–321 (Peacock and Cauchy), 321–330 (Duncan Gregory), 331–361 (De Morgan), 362–441 (Hamilton), 442–509 (Boole), 510–522 (Sylvester), 542–573 (Cayley), 362 (B. Peirce and his 162 algebras), 574–648 (C. S. Peirce). For a helpful overview of these figures, see HM, 575–582, Bourbaki, *Elements of History of Mathematics*, 117–123, and Yaglom, *Klein and Lie*, 71–94.

Grassmann: For a translation of part of his *Ausdehnungslehre* (1862), see Desmond Fearnly-Sander, "Grassmann's Theory of Dimension," in *First Australian Conference on the History of Mathematics*, 52–82; in the same volume, Fearnly-Sander also has a helpful paper on "Hermann Grassmann and the Prehistory of Universal Algebra," 41–51, which includes the statement quoted from Grassmann about the two possibilities for products of two factors, commuting and anticommuting (49).

Hamilton and the quintic: Thomas L. Hankins, *Sir William Rowan Hamilton* (Baltimore: Johns Hopkins University Press, 1980), 276–279, 248–252 (Hamilton and Peacock), 277–278 (Jerrard's faulty solution). For Hamilton's long account of Abel's argument, "On the Argument of Abel, Respecting the Impossibility of Expressing a Root of Any General Equation above the Fourth Degree, by any Finite Combination of Radicals and Rational Functions" (1839), see *The Mathematical Papers of Sir William Rowan Hamilton* (Cambridge: Cambridge University Press, 1967), 3:517–569; for Hamilton's "Inquiry into the Validity of a Method Recently Proposed by George B. Jerrard" (1837), see 3:481–516.

Jerrard: See George B. Jerrard, *Mathematical Researches* (London: Longmans, 1834).

Bolyai on quintics: See Yaglom, *Klein and Lie*, 59.

Hamilton on quaternions: HM, 582–584 and Hankins, *Hamilton*, 283–325, 247 (parents), who notes that "possibly Hamilton had been prepared for the rejection of the commutative law by discussions with Eisenstein, or possibly his attempts to find a geometrical representation of triplets had shown him that rotations in three-dimensional space do not commute either, or, what is even more likely, sacrificing commutativity seemed to him to be the only way to get any results at all" (300).

Felix Klein: For a good overview of his work, especially the Erlangen program, see Yaglom, *Klein and Lie*, 111–137. His famous 1872 Erlangen lecture is translated as "A Comparative Review of Recent Researches in Geometry," *Bulletin of the American Mathematical Society*, 2 [series 1], 215–249 (1893). See also his 1895 Göttingen lecture "The Arithmetizing of Mathematics," *Bulletin of the American Mathematical Society*, 2 [series 1], 241–249 (1896).

Duplication of the cube, trisection of angles, squaring circles: See Klein, *Famous Problems of Elementary Geometry*, 5–23.

Galois theory as "relativity": See Hermann Weyl, *Philosophy of Mathematics and Natural Science* (Princeton: Princeton University Press, 1949), 74 (Galois theory and relativity) and his *Symmetry*, 137–137 (Galois's letter).

Quantum theory and noncommutativity: See Hermann Weyl, *Symmetry*, 133–135, and *The Theory of Groups and Quantum Mechanics* (New York: Dover, 1950), 94–98. For Max Planck's view of irreversibility, see his *Eight Lectures on Theoretical Physics*, ed. Peter Pesic (Mineola, NY: Dover, 1998), vii–xiii, 1–20. I discuss this question further in my book *Seeing Double: Shared Identities in Physics, Philosophy, and Literature* (Cambridge, MA: MIT Press, 2002), 101–120.

Causality and noncommutativity: In order to obey the principle that no causal influence can travel faster than the speed of light, relativity theory imposes the constraint that quantum fields must commute when the points in question are separated by a spacelike interval, which requires greater-than-light-speed travel between them.

Nonabelian gauge theories: See Michio Kaku, *Quantum Field Theory: A Modern Introduction* (New York: Oxford University Press, 1993), 50–54 (commutation relations of the Lorentz group), 295–406 (gauge theories and the standard model). For a popular introduction, see Brian Greene, *The Elegant Universe* (New York: W. W. Norton, 1999).

Noncommutative geometry: See the authoritative (and highly technical) description in Alain Connes, *Non-Commutative Geometry* (San Diego: Academic Press, 1994). See also Pierre Cartier, "A Mad Day's Work: From Grothendieck to Connes and Kontsevich: The Evolution of Concepts of Space and Symmetry," *Bulletin of the American Mathematical Society*, 38, 389–408 (2001), which ends with speculations connecting the "cosmic Galois group" with the group of general relativity and the fine structure constant, $\alpha = e^2/\hbar c \approx 1/137$, which sets the scale for electromagnetic and quantum interactions.

Groups and space-time: I have discussed this matter in my paper "Euclidean Hyperspace and Its Physical Significance," *Nuovo Cimento* 108B, 1145–1153 (1993).

# 10  Solving the Unsolvable

De Moivre's formula: See Smith, *Source Book in Mathematics*, 440–454, including Euler's later statements of the formula. See also Eli Maor, *Trigonometric Delights* (Princeton: Princeton University Press, 1998).

Solvability in the problems of David Hilbert and Steve Smale: See Jeremy Gray, *The Hilbert Challenge* (Oxford: Oxford University Press, 2000), which includes the text of Hilbert's 1900 address (240–282), and Steve Smale, "Mathematical Problems for the Next Century," *Mathematical Intelligencer* 20:2, 7–15 (1998).

Gödel: For a thoughtful commentary on Gödel that raises the possible analogy with Abel and Ruffini, see S. G. Shanker, "Wittgenstein's Remarks on the Significance of Gödel's Theorem," in *Gödel's Theorem in Focus*, ed. S. G. Shanker (London: Routledge, 1988), 155–256 at 161–168. Less well known than Gödel's 1931 paper but of great interest is the 1936 paper by Gerhard Gentzen, "The Consistency of the Simple Theory of Types," in *The Collected Papers of Gerhard Gentzen*, ed. M. E. Szabo (Amsterdam: North-Holland, 1969), 214–222, which established that arithmetic is consistent if transfinite induction is allowed.

Ultraradical numbers: This is the term used by Stewart, *Galois Theory*, 148. D. Morduhai-Boltovski calls "hypertranscendental numbers" those that are the solutions of ordinary differential equations with constant integral coefficients, as opposed to "simply transcendental numbers," which are not the solution of any finite ordinary polynomial equation. See A. O. Gel'fond, *Transcendental and Algebraic Numbers*, tr. Leo F. Boron (New York: Dover, 1960), 96. However, the term *ultraradical* refers only to *algebraic* numbers not expressible in radicals.

Solution of the quintic: Today students have powerful, easily available computer software, such as *Mathematica*™, that is capable of dealing with solutions to the quintic. See the helpful website at http://library.wolfram.com/examples/quintic/, which is available also as a wonderful poster with all kinds of information about the history and solution of quintics. Progress continues to be made even without computers. For instance, Blair K. Spearman and Kenneth S. Williams were able to give a simple criterion in their paper "Characterization of Solvable Quintics $x^5 + ax + b$," *American Mathematical Monthly* 101, 986–992 (1994).

Computer solutions of equations: See Herman H. Goldstine, *The Computer from Pascal to von Neumann* (Princeton: Princeton University Press, 1972), 106 (Torres Quevedo).

Torres Quevedo: See Francisco González de Posada, "Leonardo Torres Quevedo," *Investigación y Ciencia*, 166:7, 80–87 (1990) and his book *Leonardo Torres Quevedo* (Madrid: Fundación Santillana, 1985).

Modular and generalized hypergeometric functions: Abel and Jacobi began to study what they called theta functions in 1827; by 1858, Charles Hermite, Leopold Kronecker, and Francesco Brioschi independently showed that any quintic could be solved by elliptic modular functions derived from these theta functions, which are infinite sums composed of powers times cosines. The notation $\sum_{n=0}^{\infty} x_n$ means the sum of the terms $x_n$ from $n = 0$ to $n = \infty$, namely $x_0 + x_1 + x_2 + \cdots$. The two theta functions are defined as

$$\theta_2(z, q) = 2 \sum_{n=0}^{\infty} q^{(n+1/2)^2} \cos((2n + 1)z),$$

$$\theta_3(z, q) = 1 + 2 \sum_{n=1}^{\infty} q^{n^2} \cos(2nz).$$

The elliptic modular function $\phi(z)$ is defined in terms of these two theta functions:

$$\phi(z) = \sqrt[3]{\frac{\theta_2(0, z)^4}{\theta_3(0, z)^4}}.$$

This function remains of central interest in contemporary mathematics. The solutions of any polynomial equation can also be expressed in terms of the generalized hypergeometric function, which is a quotient of general products of series of powers. Here the notation for products is used: $\prod_{n=1}^{p} x_n$

means the product of the terms $x_n$ from $n = 1$ to $n = p$, namely $x_1 \times x_2 \times x_3 \times \cdots \times x_p$. The generalized hypergeometric function can be expressed in this very compressed notation:

$$_pF_q(a_1, \ldots, a_p : b_1, \ldots, b_q : z) = \sum_{k=0}^{\infty} \frac{\prod_{i=1}^{p}(a_i)_k z^k}{\prod_{i=1}^{q}(b_i)_k k!},$$

where the "Pochhammer symbol" $(a_i)_k$ is defined by

$$(a_i)_k = \prod_{j=1}^{k}(a_i + j - 1).$$

All the elementary functions (such as the trigonometric functions) can be defined in terms of these extremely general functions; there are brief summaries at http://library.wolfram.com/examples/quintic/hypergeo.html and http://library.wolfram.com/examples/quintic/theta.html.

Transcendentality of $e$ and $\pi$: See the superb account of the proofs in Klein, *Famous Problems of Elementary Geometry*, 61–77. For a general history, see Petr Beckmann, *A History of $\pi$*, second edition (Boulder, CO; Golem Press, 1971); for a valuable collection of original papers, see *Pi: A Source Book*, ed. Lennart Berggren, Jonathan Borwein, and Peter Borwein (New York: Springer, 1997), which includes Lindemann's original proof (194–229), and Ivan Niven, "A Simple Proof That $\pi$ is Irrational," 509.

Cantor's proof: For a translation of his crucial paper, see Ewald, *From Kant to Hilbert*, 2:838–940. There is a nice brief account in Klein, *Famous Problems of Elementary Geometry*, 49–55. Dunham gives a helpful and accessible account of Cantor's work in his *Journey through Genius*, 245–283. For a detailed study of the development of Cantor's ideas, see Joseph Warren Dauben, *Georg Cantor: His Mathematics and Philosophy of the Infinite* (Princeton: Princeton University Press, 1990).

Concepts of infinity: For Richard Dedekind's introduction of the infinite as a fundamental notion of set theory, see his *Essays on the Theory of Numbers* (New York: Dover, 1963), 63–64. For a general overview, see Eli Maor, *To Infinity and Beyond: A Cultural History of the Infinite* (Boston: Birkhäuser, 1987). For the connection with art, see J. V. Field, *The Invention of Infinity: Mathematics and Art in the Renaissance* (Oxford: Oxford University Press, 1997).

Beauty in mathematics: For one possible realization of "classic" versus "romantic" mathematical styles, see François Le Lionnais, "Beauty in Mathematics," in his collection *Great Currents of Mathematical Thought* (New York: Dover, 1971), 2:121–158.

Kant: For his account of the "mathematical sublime," see Immanuel Kant, *Critique of the Power of Judgement*, tr. Paul Guyer and Eric Matthews (Cambridge: Cambridge University Press, 2000), 131–149.

Abel's work on topics besides solvability of equations: See Michael Rosen, "Abel's Theorem on the Lemniscate," *American Mathematical Monthly* 88, 387–394 (1981), and (concerning Abel's contributions on elliptic integrals) Roger Cooke, "Abel's Theorem," in *The History of Modern Mathematics*, ed. David E. Rowe and John McCleary (New York: Academic Press, 1989), 389–421. For Abel's summary comments on the lemniscate, see his 1826 letter to Holmboe, cited in *Œuvres Complètes de Niels Henrik Abel*, 2:261–262.

Abel's "favorite subject": See his 1826 letter to Holmboe, cited in *Œuvres Complètes de Niels Henrik Abel*, 2:260.

Abel's credo: My translation of part of his introduction to "Sur la Resolution Algébrique des Équations," *Œuvres Complètes de Niels Henrik Abel*, 2:217–243 at 217; the editors comment (329–338) that this paper was first published by Holmboe in 1839 and also give interesting alternative readings of Abel's text.

Abel's Paris notebooks: Stubhaug, *Abel*, 504–505, reproduces and translates this page (504), along with another from this notebook. Though apparently unmoved by music, Abel was very fond of the theater. During his travels, he may well have seen Beaumarchais's popular play *Le Mariage de Figaro*, which contains a virtuoso scene (III.v) in which Figaro plays with the English expression "God-dam" (as Abel also spells it).

## Appendix A    Abel's 1824 Paper

This is Abel's original version of his proof, in my translation. The original text is in his *Œuvres*, 1:28–33. The only other published translation, in Smith, *Source Book in Mathematics*, 261–266, unfortunately has several significant misprints. My notes try to explain Abel's brief indications, which he expanded in his 1826 version of the proof, "Démonstration de l'impossibilité de la résolution algébrique des équations générales qui passent le quatrième degré," *Œuvres*, 1:66–94.

## Appendix B    Abel on the General Form of an Algebraic Solution

My account is drawn from Abel's "Démonstration de l'impossibilité," 66–72.

## Appendix C    Cauchy's Theorem on Permutations

The original source is Cauchy's "Mémoire sur le nombre des valeurs qu'une fonction peut acquérir" (1815), in *Œuvres Complète d'Augustin Cauchy* (Paris: Gauthier-Villars, 1905), series II, 1:62–90. My account follows closely Abel's account in his "Démonstration de l'impossibilité," 75–79.

# Acknowledgments

To Larry Cohen and his associates at the MIT Press for their support and collaboration in bringing this book to life.

To St. John's College for released time under the Louise Trigg tutorship, and to my fellow students who encouraged my struggle to understand Abel's proof.

To Raymond Ayoub, David Cox, David Derbes, William Dunham, Robin Hartshorne, Barry Mazur, Mark Peterson, Michael Rosen, Tony Rothman, Jerry Shurman, and Curtis Wilson, whose comments and criticisms saved me from many mistakes. The errors that remain are my own.

And to Ssu, Andrei, and Alexei, who solved the unsolvable for me.

# Index

Abbati, Pietro, 82
Abel, Niels Henrik, 1–3, 85–102,
    189n–190n
  Abelian Addition Theorem,
    151, 200n
  abelian equations, 98–100
  abelian groups, 112–113
  Abelian integrals, 151, 200n
  Abel-Ruffini Theorem, 1–3, 89–94,
    155–170, 200n
  and Cauchy, 87, 93–94, 96
  early life, 85–89, 189n
  formulas of, 88
  and Galois, 105, 108–109,
    130–131, 145
  and Gauss, 88–89, 95, 151
  and Hamilton, 133
  illness and death of, 101–102, 190n
  notebooks of, 97, 152–153, 200n
  and Ruffini, 87–89, 97, 190n
  travels in Europe, 95–97
Académie des Sciences, 96, 104–106
Accounting. See Bookkeeping
Algebra
  Arabic, 23–28, 45, 54
  coefficient, 1, 44–45, 91–93
  noncommutative, 131–143
  roots, 1, 92, 98

symbolic notation, 40–45, 132
unknown, 43
variable, 1, 44
Algebraic functions, 90–91
al-Khwārizmī, Muhammed
    ibn-Musa, 25–26, 30, 183n–184n
Alogon, 9
Alternating groups. See Groups
Analytic mathematics, 42, 59
Anrta, 9
Anticommutation, 134–136, 195n
Apollonius of Perga, 42, 51, 59
Aporia, 14, 182n
Archimedes, 32, 60
Area problem, 61–66
Aristotle, 7, 181n
Arithmos, 9, 181n
Athens, 15
Ausdehnungslehre (Grassmann),
    135–136
Ayoub, Raymond, 189n, 192n–193n

Babylonian mathematics, 5, 7,
    24–25, 30, 183n
Bacon, Francis, 182n–183n
Basham, A. L., 181n
Beaumarchais, Pierre Augustin
    de, 200n

Bell, Eric Temple, 191n
Bernoulli, Daniel, 65
Bolyai, János, 133, 196n
Bombelli, Raphael, 54–55, 187n
Bookkeeping, double-entry, 27–29,
    184n–185n
Boole, George, 132–133, 195n
  Boolean algebra, 132–133
Born, Max, 141
Bourbaki, Nicolas (pseudonym),
    188n, 195n
Boyer, Carl, 182n–195n
Bring, E. S., 67, 188n
Brioschi, Fernando, 146, 198n
Brizio, Anna Maria, 184n
Brown, R. G., 184n
Bryce, R. A., 189n
Bühler, W. K., 189n
*Bulletin* (Baron de Férussac), 96
Bürgi, Jost, 48–49
Burkert, Walter, 1981n
Burn, R. P., 193n
Burnside, William Snow, 190n,
    192n–193n

Cajori, Florian, 186n
Calculators, 78, 149–150
Calculus, 60, 62–63
Cantor, Georg, 150, 199n
Cardano, Girolamo, 30–40, 54, 57,
    67, 69, 185n
Cartier, Pierre, 197n
Cauchy, Augustin, 83, 87, 96,
    104–105
  Cauchy's theorem, 93, 163, 166,
    175–180, 201n
  and commutativity, 132, 195n
Causality and noncommutativity,
    142, 196n
Cayley, Arthur, 112, 126,
    136–138, 195n
  Cayley numbers, 137

Cayley tables, 112,
    119, 122
Cervantes, Miguel de, 23
Charles X, 105
Chatfield, Michael, 185n
Chinese mathematics, 183n
Christiania. *See* Oslo
Cipher, 27–28, 33, 42
Circle, 62
Code. *See* Cipher
Coleridge, Samuel Taylor, 135
Commensurable, 7–8, 16
Commercial arithmetic, 27–31
Commutativity, 99, 112
"Completing the cube," 36–37, 120
"Completing the square,"
    25–26, 113
Computers, 147, 197n–198n
Conic sections, 42
Connes, Alain, 197n
Continuum, 11, 18
Cooke, Roger, 200n
*Cosa* (coss), 27, 44
Crelle, August, 95–97, 100, 102
  *Crelle's Journal*, 95, 101
Cross product. *See* Multiplication
Cube, 5–6
  symmetries of, 121
Cyclic groups and symmetries, 113,
    117, 122, 195n

d'Alembert, Jean Le Rond, 68
Dance, 111–130
Dauben, Joseph Warren, 199n
Dedekind, Richard, 183n, 199n
Dehn, Edgar, 192n
del Ferro, Scipione, 32–34
del Ferro–Cardano–Tartaglia
    method, 32–35, 48–49, 54–55,
    77, 174
De Moivre, Abraham, 149, 197n
DeMorgan, Augustus, 132, 195n

Descartes, René, 50–59, 68, 187n
  and conic sections, 64
  Descartes's rule of signs, 53
  *La Geometrie,* 50–58
  "relativity" of roots, 57, 140
Dickson, Leonard E., 192n
Dimension (algebraic), 50–51
*Disquisitiones Arithmeticae* (Gauss), 79, 189n
Dodecahedron, 5–6
  symmetries of, 124–125, 126–130
*Don Quixote* (Cervantes), 23
Dörrie, Heinrich, 188n–190n
Dunham, William, 185n, 199n
Duplication of cube, 196n

*e,* 28, 150, 184n, 199n
École Polytechnique, 105
École Préparatoire (École Normale Superieure), 105
Edwards, Harold M., 191n
Einstein, Albert, 140, 143
Eisenstein, E. L., 184n
*Elements. See* Euclid
Equations, algebraic
  approximate solutions, 66, 147, 198n
  cubic, 3, 28, 30–37, 90, 113–120, 148–149, 185n
  general formulation, 1–3
  quadratic, 2, 23, 25–26, 64, 90–91, 111–113, 185n
  quartic, 2, 35, 38–39, 76–78, 120–122
  quintic, 2–3, 77–78, 91–99, 122–129, 198n
  roots, 1
Erlangen Program (Felix Klein), 138–140, 196n
Euclid, 5, 17–23, 42, 59, 145–146, 150, 183n
Euclidian geometry, 139

Eudoxus, 17–18, 183m
Euler, Leonhard, 62, 68, 90, 149, 196n
Ewald, William B., 187n, 195n, 199n

Fauvel, John, 184n–185n
Fearnly-Sander, Desmond, 195n
Fermat's Last Theorem, 87–88, 189n–190n
Ferrari, Ludovico (Luigi), 34–35, 37–39, 57, 69, 76, 122
Fibonacci. *See* Leonardo of Pisa
Field, J. V., 186n, 199n
Fields (mathematics), 139
Fields (physics), quantum theory of, 142–143
Fine, Benjamin, 188n
Fine structure constant, 197n
Fontana, Niccolò. *See* Tartaglia
Fractions, 7
France, 45, 96–97, 102–108, 190n
Fundamental Theorem of Algebra, 56, 68–73, 79, 146, 188n

Galilei, Galileo, 49–50, 187n
Galois, Évariste, 102–109, 190n–191n
  and Abel, 105–106, 108–109, 130–131, 145
  and Cauchy, 104–105
  death of, 106–108
  education of, 102–106, 190n
  and his father, 104–105
  Galois theory, 125–130, 191n–193n
  legend of, 108, 191n
  posthumous writings of, 108
  and Société des Amis du Peuple, 106–107
  and Stéphanie Poterin-Dumotel, 106
Gårding, Lars, 190n

Gauge fields, nonabelian, 142–143, 196n
Gauss, Carl Friedrich, 70–74, 97, 187n–189n
and Abel, 89, 95, 100, 151
and commutativity, 131–132, 195n
and unsolvability of quintic, 79, 88
Gazalé, Midhat, 183n
Gel'fond, A. O., 197n
Gentzen, Gerhard, 197n
*Geometrie, La* (Descartes), 50–54, 187n
Geometry, 50, 60, 66
Germain, Sophie, 104
Gibbs, Josiah Willard, 136
Gibbs, W. Wayt, 195n
Gies, J. and F., 184n
Girard, Albert, 51, 56, 68, 187n
Girard's identities, 61, 92
Gleason, Andrew, 187n
God, 49, 55
Gödel, Kurt, 197n
"Golden ratio," 28
Goldstine, Herman H., 198n
González de Posada, Francisco, 198n
Gorman, Peter, 181n
Grafton, Anthony, 185n
Grassmann, Hermann, 135–136, 195n
Gray, Jeremy, 184n–185n, 197n
*Great Art* (Cardano), 30–40, 185n
Greek mathematics, 5–21
Greene, Brian, 196n
Gregory, Duncan, 132, 195n
Grossmann, Israel, 193n
Groups, 109, 111–130, 138–140, 193n–195n
$A_3$, 118–120
$A_4$, 121–122
$A_5$, 123–129, 139
abelian, 112–113, 129

continuous, 140
cosets, 176
cyclical, 113, 117, 122, 175–180, 195n
definition of, 125–126
identity, 112, 119, 125
invariant subgroups, 119, 129
Lagrange's Theorem, 128, 175–176
Lorentz, 196n
monster group, 130, 195n
nonabelian, 118, 129, 142–143
normal subgroups, 119, 129, 193n–195n
order, 176
and permutations, 175–180
philosophical aspects, 193n
quotient, 130, 176–177, 193n–194n
$S_2$, 112–113
$S_3$, 113–120, 139
$S_4$, 120–122, 139
$S_5$, 122–124
simple groups, 130
solvable chains of, 130, 194n–195n
$V$, 122
visualization of, 193n
Guérard, Albert, 190n

Hadlock, Charles Robert, 192n
Hamilton, William Rowan, 133–136, 196n
Hankins, Thomas L., 196n
*Harmony of the World* (Kepler), 48, 121, 124, 186n
Hartshorne, Robin, 181n, 191n, 194n
Heath, Thomas, 183n
Heaviside, Oliver, 136
Heisenberg uncertainty principle, 141
Hellman, Morton J., 185n
Henry IV, 45
Heptagon, 48, 186n–187n

Hermite, Charles, 146, 150, 198n
Herrstein, I. N., 192n
Hexagon, 48
Hilbert, David, 197n
Hippias of Mesopontum, 10
Hirano, Yoïchi, 193n
Hoe, J., 183n
Hölder, Otto, 130–131, 177, 194n
Holmboe, Berndt Michael, 87, 97,
    190n, 200n
Holy Spirit, 55
Huffman, C. A., 181n
Huntley, H. E., 181n
Hypergeometric functions,
    198n–199n

Icosahedron, 5–6
  symmetries of, 123–129
Incommensurability, 7–14
Indian mathematics, 9
Indistinguishability of quanta, 142
Infinity, 22, 146, 148, 151, 153
Institut de France, 100, 106
Invariance, 113, 139
Invariant subgroups. See Groups
Irrational magnitudes, 7–14, 19–21,
    145–146, 183n
Irreducible case (cubic
    equations), 54
Irreversibility, 141
Isograph, 147

Jacobi, Carl Gustav Jacob,
    100, 146
Jacobson, Nathan, 192n
Jerrard, George B., 67, 133,
    188n, 195n
Johnston, K. S., 184n
Jordan, Camille, 130–131, 133, 146,
    177, 194n
Jordan, Pascual, 141
"July monarchy," 105–106

Kabbalists, 48
Kaku, Michio, 196n
Kant, Immanuel, 200n
Karl XIII, 85
Kemp, Christine, 96, 101–102
Kepler, Johannes, 48–49, 121, 124,
    186n–187n
Khayyām, Omar, 30, 184n
Kiernan, B. Melvin, 193n
King, R. Bruce, 192n
Klein, Felix, 138–140, 143, 189n,
    191n, 196n, 199n
Klein, Jacob, 182n, 187n
Kline, Morris, 182n, 192n
Knorr, Wilbur Richard, 182n
Kronecker, Leopold, 146, 190n, 198n

Lafayette, General, 105
La Geometrie (Descartes), 50–58
Lagrange, Joseph-Louis, 73–83,
    87, 188n
  Lagrange resolvent, 74–79
  Lagrange's Theorem, 128,
    175–176, 194n
Lalanne, Leon, 147
La Nave, Federica, 187n
Laplace, Pierre Simon, 51, 80
Legendre, Adrien-Marie, 96,
    100–101
Leibniz, Gottfried Wilhelm, 55,
    65–67, 183n, 187n–188n
Le Lionnais, François, 199n
Le Mariage de Figaro
    (Beaumarchais), 200n
Lemniscate, 65, 152–153, 200n
Leonardo da Vinci, 6, 28, 184n
Leonardo of Pisa (Fibonacci), 27–28,
    30, 184n
Lieber, Lillian R., 192n
Lindemann, Ferdinand, 150, 199n
Liouville, Joseph, 133
Littlewood, D. E., 192n

Locus problem, 57, 59
*Logos,* 9
Louis XVI, 104
Louis XVIII, 104–105
Louis-Philippe I, 105–106
Lycée Louis-le-Grand, 104

Macve, Richard, 184n
Magnitudes, 7–8, 23
Magnus, Wilhelm, 193n
Malfatti, Gianfrancesco, 77, 82
Maor, Eli, 184n, 197n, 199n
Marinoni, Augusto, 184n
*Mathematica*[TM], 198n
Matrix, 136–138
Maxfield, John E. and Margaret W.,
    191n–192n
Maxwell, James Clerk, 135–136
Maxwellian dynamics, 141
Mayer, Uwe F., 188n
Mazur, Barry, 187n
Meno, 13–14, 182n
*Mercantile Arithmetic* (Widman), 29
Merzbach, Uta C., 182n–195n
Minkowski, Hermann, 140
Mitchell, David, 193n
Modular functions, 198n
Monster. *See* Groups
Montucla, Jean Étienne, 79, 189n
Morduhai-Boltovsky, D., 197n
Multiplication
  commutativity of, 131–132
  Grassmann algebra, 135–136
  matrix, 136–138
  quaternion, 134
  scalar product, 135
  vector product, 135
Music, 7, 19–20, 48, 183n
Mutafian, Claude, 192n

Nahin, Paul J., 187n
Napoleon, 104, 108

Needham, Joseph, 183n
Newton, Isaac, 59–66, 149–150,
    187n–188n
  and Descartes, 59, 66
  lemma 28, 61–66, 148
  Newton's identities, 60–61
  Newton's method, 66
Newtonian dynamics,
    136, 141
Niven, Ivan, 199n
Nonabelian gauge fields,
    142–143, 196n
Nonabelian groups. *See* Groups
Noncommutative geometry,
    143, 197n
Noncommutativity, 99–100,
    131–143, 195n
Normal subgroups. *See* Groups
Norway, 85
Numbers
  algebraic, 146, 150, 197n
  complex and imaginary, 54–56, 70,
    148–149, 187n
  counting, 9
  in Greek mathematics, 9
  irrational magnitudes, 7–8, 18–19,
    23, 146
  line, 51
  negative, 51–54, 187n
  octonions (Cayley numbers), 137
  place value, 24
  quaternions, 134–135, 196n
  rational, 7–8, 146
  sexagesimal, 24–25
  transcendental, 62, 66, 150,
    197n, 199n
  ultraradical, 146, 150, 197n

Octahedron, 5–6
  symmetries of, 121
Octonions. *See* numbers
Ore, Øystein, 185n, 189n

Oslo, 87, 101
Oval, 61–66

Pacioli, Luca, 6, 28–30, 184n–185n
Panton, Arthur William,
    190n, 192n
Pappus, 10–11, 42, 57, 182n
Parabola, 65
Parshall, Karen Hunger, 184n–185n
Pascal, Blaise, 147
Peacock, George, 132, 195n
Pentagon, 49
Permutations, 75–77, 82, 108–109,
    111–130, 175–180
Pesic, Peter, 142, 182n–183n, 186n,
    188n, 196n–197n
Peterson, Mark, 185n
Pi ($\pi$), 62, 150, 199n
Pierce, Benjamin, 138, 195n
Pierce, C. S., 138, 195n
Piero della Francesca, 28, 30, 185n
Pierpont, J., 190n
Planck, Max, 141, 196n
Plato, 11–17, 44, 140–141, 182n
Platonic solids, 5–6, 122, 138,
    143, 193n
Poisson, Siméon–Denis, 101
Postnikov, M. M., 192n
Poterin-Dumotel, Stéphanie, 106
Pourciau, Bruce, 188n
Principia (Newton), 59–66, 187n
Pycior, Helena M., 186n–187n
Pythagoras, 5–11, 46, 181n
  Pythagorean theorem, 11
Pythagoreans, 5–11, 15

Quantum theory, 141–143, 196
Quaternions. See numbers

Radicals, 2, 35
Ralph, Leslie, 181n
Rashed, Roshdi, 184n

Raspail, François-Vincent, 97, 106,
    108, 191n
Rational magnitudes, 9
Ratios, 7
Reductio ad absurdum, 7–8, 64, 90
Relativity
  and Galois theory, 140, 196n–197n
  general, 143, 196n
  of roots, 57, 140
  special, 140
  of space-time, 140
Republic (Plato), 15, 182n
Resolvent, see Lagrange resolvent
Richard, Louis-Paul-Émile, 104–105
Roman law, 43
Roots of unity, 74, 97
Rosen, Michael, 190n, 200n
Rosenberger, Gerhard, 188n
Rothman, Tony, 191n
Royal Frederick's University,
    Christiania (Oslo), 87
Rta, 9, 181n
Ruffini, Paolo, 80–83

Sacrifice, 10, 46
Saigey, Jaques Frédéric, 96
Scalars, 135
Second Law of
    Thermodynamics, 141
Seventeen-sided polygon, 70,
    74, 189n
Shanker, S. G., 197n
Shurman, Jerry, 192n
Shylock, 27, 184n
Singh, Simon, 190n
Skau, Christian, 190n
Smale, Steve, 197n
Société des Amis du Peuple,
    106–107
Socrates, 13–17, 182n
Solomon, Ron, 195n
Solution in radicals, 2

Space
    four-dimensional 135, 197n
    *n*-dimensional, 135–136, 138
    three-dimensional, 139–141, 143
Spearman, Blair K., 198n
Species, logic of, 44, 132
Speed of light, 140, 142, 196n
Square, 7–14
Square roots, sound of, 20
Squaring the circle, 150, 196n
Stahl, Saul, 191n, 195n
"Standard theory" (physics),
    142, 196n
Stein, Howard, 183n
Steiner, George, 191n
Stewart, Ian, 192n, 197n
Stillwell, John, 193n
Stubhaug, Arild, 189n–191n, 200n
Subgroups. *See* Groups
Suleiman II, 153
*Summary of Arithmetic* (Pacioli),
    28–29
Swetz, Frank, 185n
Sylvester, James Joseph,
    136–138, 195n
Symmetric groups. *See* Groups
Symmetry, 113
    in algebraic expressions, 60
    of fundamental particles, 142–143
    of polyhedra (*see* Triangle;
        Tetrahedron; Cube;
        Dodecahedron; Icosahedron;
        Octahedron)
    of three-dimensional space, 124
Synthetic mathematics, 42–43

Tartaglia, 32–34, 185n
Taton, René, 191n
Taylor, R. Emmett, 184n
Tetractys, 10
Tetrahedron, 5–6
    symmetries of, 120–121

Theaetetus, 15–17, 20, 182n
Theodorus, 17, 182n
Theology, Christian, 55
Thermodynamics, 141
Theta functions, 146, 198n
Third Law of Planetary Motion
    (Kepler), 49
Tignol, Jean–Pierre, 192n
Time, irreversibility of, 141
Topology, 72
Torres Quevedo, Leonardo,
    147, 198n
Torture, 16–17, 182n–183n
Toti Rigatelli, Laura, 190n–192n
Transcendental. *See* Number
"Treviso arithmetic," 29
Trial, 18
Triangle, symmetries of, 113–120
Trigonometry, 47, 149–150, 186n
Trisection of angle, 187n, 196n
Tschebotaröw, N., 192n

*Universal Arithmetic* (Newton), 60
Uspensky, J. V., 185n, 188n

Vandermonde, Alexandre-
    Théophile, 75, 188n
van der Waerden, B. L., 183n–184n,
    188n, 193n
van Roomen, Adriaan, 45–47, 186n
Vectors, 135
Vernier, Hippolyte Jean, 104
Verriest, G., 192n
Vetter, Guido, 186n
Viète, François, 41–47, 56–58, 73,
    132, 151, 186n
Voltaire, François Marie
    Arouet de, 85
von Fritz, Kurt, 182n
von Tschirnhaus, Count Ehrenfried
    Walter, 66–68, 188n
    Tschirnhaus transformation, 68

Walker, D. P., 186n
Wallis, John, 186n
Weil, Simone, 183n
West, M. L., 183n
Weyl, Hermann, 140, 193n, 196n
Widman, Johann, 29
Wiles, Andrew, 189n
Williams, Kenneth S., 198n
Wilson, Curtis, 193n
Winternitz, Emmanuel, 184n
Wussing, Hans, 188n–189n, 193n

Xenophon, 182n

Yaglom, I. M., 192n, 195n–196n
Yamey, B. S., 184n

Zammattio, Carlo, 184n
Zero, 9, 43